Symmetry in Mathematical Analysis and Application

Symmetry in Mathematical Analysis and Application

Special Issue Editor
Luigi Rodino

MDPI • Basel • Beijing • Wuhan • Barcelona • Belgrade

Special Issue Editor
Luigi Rodino
University of Torino
Italy

Editorial Office
MDPI
St. Alban-Anlage 66
4052 Basel, Switzerland

This is a reprint of articles from the Special Issue published online in the open access journal *Symmetry* (ISSN 2073-8994) from 2018 to 2019 (available at: https://www.mdpi.com/journal/symmetry/special_issues/Symmetry_Mathematical_Analysis_Applications).

For citation purposes, cite each article independently as indicated on the article page online and as indicated below:

LastName, A.A.; LastName, B.B.; LastName, C.C. Article Title. *Journal Name* **Year**, *Article Number*, Page Range.

ISBN 978-3-03936-411-4 (Hbk)
ISBN 978-3-03936-412-1 (PDF)

© 2020 by the authors. Articles in this book are Open Access and distributed under the Creative Commons Attribution (CC BY) license, which allows users to download, copy and build upon published articles, as long as the author and publisher are properly credited, which ensures maximum dissemination and a wider impact of our publications.

The book as a whole is distributed by MDPI under the terms and conditions of the Creative Commons license CC BY-NC-ND.

Contents

About the Special Issue Editor . . . . **vii**

Preface to "Symmetry in Mathematical Analysis and Application" . . . . **ix**

Arnon Ploymukda and Pattrawut Chansangiam
Integral Inequalities of Chebyshev Type for Continuous Fields of Hermitian Operators Involving Tracy–Singh Products and Weighted Pythagorean Means
Reprinted from: *Symmetry* **2019**, *11*, 1256, doi:10.3390/sym11101256 . . . . **1**

Paweł Więcek, Daniel Kubek, Jan Hipolit Aleksandrowicz and Aleksandra Stróżek
Framework for Onboard Bus Comfort Level Predictions Using the Markov Chain Concept
Reprinted from: *Symmetry* **2019**, *11*, 755, doi:10.3390/sym11060755 . . . . **13**

Manuel De la Sen
On Cauchy's Interlacing Theorem and the Stability of a Class of Linear Discrete Aggregation Models Under Eventual Linear Output Feedback Controls
Reprinted from: *Symmetry* **2019**, *11*, 712, doi:10.3390/sym11050712 . . . . **26**

S. A. Alharbi, A. S. Rambely
A Dynamic Simulation of the Immune System Response to Inhibit And Eliminate Abnormal Cells
Reprinted from: *Symmetry* **2019**, *11*, 572, doi:10.3390/sym11040572 . . . . **48**

Astha Chauhan, Rajan Arora and Mohd Junaid Siddiqui
Propagation of Blast Waves in a Non-Ideal Magnetogasdynamics
Reprinted from: *Symmetry* **2019**, *11*, 458, doi:10.3390/sym11040458 . . . . **62**

Jawdat Alebraheem
Relationship between the Paradox of Enrichment and the Dynamics of Persistence and Extinction in Prey-Predator Systems
Reprinted from: *Symmetry* **2018**, *10*, 532, doi:10.3390/sym10100532 . . . . **75**

About the Special Issue Editor

Luigi Rodino is, at present, has been a professor at the University of Torino, Italy, since 1976. He completed his degree in Mathematics at the University of Torino in 1971, and his post-doctoral studies 1972–75, which he completed at the University of Lund in Sweden, Institut Mittag Leffler in Sweden, and the University of Princeton, USA. Previously, at the University of Torino, he was the Director of the Department of Mathematics 1988–91, and the President of the Faculty in Mathematics for Finance and Insurance 2006–2009. He was also the Coordinator of International Research Projects for NATO and UNESCO 1995–2005. Moreover, he was the President of ISAAC (International Society Analysis Applications Computations), 2013–16. He is currently the Editor-in-Chief of two international journals, and a member of the Editorial Committee for 22 journals. His main research fields are partial differential equations and Fourier analysis, and he is the author of 145 papers, 6 monographies, 16 edited volumes and 800 reviews. He has supervised 17 Ph.D. students so far.

Preface to "Symmetry in Mathematical Analysis and Application"

'Mathematics servants of Sciences, Mathematics queens of Sciences.' This is the rough translation of a statement in Latin, describing the role of mathematics in the scientific community. At the core of mathematics, mathematical analysis in the past centuries has provided applications in different disciplines that are essential for accessing modern knowledge, in both practical and theoretical aspects. In addition to these applications, mathematics possesses a wonderful beauty: fundamental formulas present deep links in symmetry which go beyond technical expressions.

This Special Issue of *Symmetry* consists of six articles devoted to models in medicine, biology, ecology and other disciplines, all expressed in terms of mathematical analysis, showing the effectiveness of mathematics in different aspects of modern life. Other contributions in pure mathematics also give evidence for the role of Symmetry in these theoretical aspects.

In this preface, the six articles in the Special Issue will be addressed, and a detailed presentation of the different topics will be provided. Here, then, we will give an idea of some of the relevant achievements in the present volume.

In medicine, biology and ecology, immune system response is studied and related to the risk of cancer. The increase of the nutrition of the prey and destabilizing the predator–prey dynamics are both considered.

In symmetry in mathematics, the classical symmetric means are generalized to weighted Pythagorean means. The eigenvalues of the sequences of matrices are studied, in connection with stability and convergence problems.

Other relevant contributions concern the efficiency of public transport, with particular reference to the reduction of congestion, energy consumption and emissions. Blast waves are also considered, in particular in the relevant case of supersonic speed, as in explosions.

Overall, the volume is an excellent report on the relevance of mathematical analysis in applied sciences, with an emphasis placed on the deep relations with *Symmetry*.

Luigi Rodino
Special Issue Editor

Article

Integral Inequalities of Chebyshev Type for Continuous Fields of Hermitian Operators Involving Tracy–Singh Products and Weighted Pythagorean Means

Arnon Ploymukda and Pattrawut Chansangiam *

Department of Mathematics, Faculty of Science, King Mongkut's Institute of Technology Ladkrabang, Bangkok 10520, Thailand; arnon.p.math@gmail.com
* Correspondence: pattrawut.ch@kmitl.ac.th; Tel.: +66-935-266600

Received: 14 August 2019; Accepted: 5 October 2019; Published: 9 October 2019

Abstract: In this paper, we establish several integral inequalities of Chebyshev type for bounded continuous fields of Hermitian operators concerning Tracy-Singh products and weighted Pythagorean means. The weighted Pythagorean means considered here are parametrization versions of three symmetric means: the arithmetic mean, the geometric mean, and the harmonic mean. Every continuous field considered here is parametrized by a locally compact Hausdorff space equipped with a finite Radon measure. Tracy-Singh product versions of the Chebyshev-Grüss inequality via oscillations are also obtained. Such integral inequalities reduce to discrete inequalities when the space is a finite space equipped with the counting measure. Moreover, our results include Chebyshev-type inequalities for tensor product of operators and Tracy-Singh/Kronecker products of matrices.

Keywords: Chebyshev inequality; Tracy-Singh product; continuous field of operators; Bochner integral; weighted Pythagorean mean

1. Introduction

One of the fundamental inequalities in mathematics is the Chebyshev inequality, named after P.L. Chebyshev, which states that

$$\frac{1}{n}\sum_{i=1}^{n} a_i b_i \geqslant \left(\frac{1}{n}\sum_{i=1}^{n} a_i\right)\left(\frac{1}{n}\sum_{i=1}^{n} b_i\right) \tag{1}$$

for all real numbers a_i, b_i ($1 \leqslant i \leqslant n$) such that $a_1 \leqslant \ldots \leqslant a_n$ and $b_1 \leqslant \ldots \leqslant b_n$, or $a_1 \geqslant \ldots \geqslant a_n$ and $b_1 \geqslant \ldots \geqslant b_n$. This inequality can be generalized to

$$\sum_{i=1}^{n} w_i a_i b_i \geqslant \left(\sum_{i=1}^{n} w_i a_i\right)\left(\sum_{i=1}^{n} w_i b_i\right) \tag{2}$$

where $w_i \geqslant 0$ for all $1 = 1, \ldots, n$. A matrix version of (2) involving the Hadamard product was obtained in [1].

A continuous version of the Chebyshev inequality [2] says that if $f, g : [a, b] \to \mathbb{R}$ are monotone functions in the same sense and $p : [a, b] \to [0, \infty)$ is an integrable function, then

$$\int_a^b p(x)dx \int_a^b p(x)f(x)g(x)dx \geqslant \int_a^b p(x)f(x)dx \cdot \int_a^b p(x)g(x)dx. \tag{3}$$

If f and g are monotone in the opposite sense, the reverse inequality holds. In [3], Moslehian and Bakherad extended this inequality to Hilbert space operators related with the Hadamard product by using the notion of synchronous Hadamard property. They also presented integral Chebyshev inequalities respecting operator means.

The Grüss inequality, first introduced by G. Grüss in 1935 [4], is a complement of the Chebyshev inequality. This inequality gives a bound of the difference between the product of the integrals and the integral of the product for two integrable functions. For each integral function $f : [a,b] \to \mathbb{R}$, let us denote

$$\mathcal{I}(f) = \frac{1}{b-a}\int_a^b f(x)dx.$$

The Grüss inequality states that if $f, g : [a, b] \to \mathbb{R}$ are integrable functions and there exist real constants k, K, l, L such that $k \leqslant f(x) \leqslant K$ and $l \leqslant g(x) \leqslant L$ for all $x \in [a, b]$, then

$$|\mathcal{I}(fg) - \mathcal{I}(f)\mathcal{I}(g)| \leqslant \frac{1}{4}(K-k)(L-l). \qquad (4)$$

This inequality has been studied and generalized by several authors; see [5–7]. In [7], the term Chebyshev-Grüss inequalities is used mentioning to Grüss inequalities for Chebyshev functions $T_\mathcal{I}$ which defined as

$$T_\mathcal{I}(f,g) = \mathcal{I}(f \cdot g) - \mathcal{I}(f) \cdot \mathcal{I}(g).$$

A general form of Chebyshev-Grüss inequalities is given by

$$|T_\mathcal{I}(f,g)| \leqslant E(\mathcal{I}, f, g)$$

where E is an expression depending on the arithmetic integral mean $\mathcal{I}$ and oscillations of f and g. Chebyshev-Grüss inequalities for some kind of operator via discrete oscillations is presented by Gonska, Raça and Rusu [7].

On the other hand, the notion of tensor product of operators is a key concept in functional analysis and its applications particularly in quantum mechanics. The theory of tensor product of operators has been investigate in the literature; see, e.g., [8,9]. In [10,11], the authors extend the notion of tensor product to the Tracy-Singh product for operators on a Hilbert space, and supply algebraic/order/analytic properties of this product.

In this paper, we establish a number of integral inequalities of Chebyshev type for continuous fields of Hermitian operators relating Tracy-singh products and weighted Pythagorean means. The Pythagorean means considered here are three classical means -the geometric mean, the arithmetic mean, and the harmonic mean. The continuous field considered here is parametrized by a locally compact Hausdorff space Ω endowed with a finite Radon measure. In Section 2, we give basic results on Tracy-Singh products for Hilbert space operators and Bochner integrability of continuous field of operators on a locally compact Housdorff space. In Section 3, we provide Chebyshev type inequalities involving Tracy-Singh products of operators under the assumption of synchronous Tracy-Singh property. In Section 4, we establish Chebyshev integral inequalities concerning operator means and Tracy-Singh products under the assumption of synchronous monotone property. Finally, we prove Chebyshev-Grüss inequalities via oscillations for continuous fields of operators in Section 5. In the case that Ω is a finite space with the counting measure, such integral inequalities reduce to discrete inequalities. Our results include Chebyshev-type inequalities concerning tensor product of operators and Tracy-Singh/Kronecker products of matrices.

2. Preliminaries

In this paper, we consider complex Hilbert spaces $\mathbb{H}$ and $\mathbb{K}$. The symbol $\mathbb{B}(\mathbb{X})$ stands to the Banach space of bounded linear operators on a Hilbert space $\mathbb{X}$. The cone of positive operators on $\mathbb{X}$ is denoted by $\mathbb{B}(\mathbb{X})^+$. For Hermitian operators A and B in $\mathbb{B}(\mathbb{X})$, the situation $A \geq B$ means that $A - B \in \mathbb{B}(\mathbb{X})^+$. Denote the set of all positive invertible operators on $\mathbb{X}$ by $\mathbb{B}(\mathbb{X})^{++}$.

We fix the following orthogonal decompositions:

$$\mathbb{H} = \bigoplus_{i=1}^{m} \mathbb{H}_i, \quad \mathbb{K} = \bigoplus_{k=1}^{n} \mathbb{K}_k$$

where all $\mathbb{H}_i$ and $\mathbb{K}_j$ are Hilbert spaces. Such decompositions lead to a unique representation for each operator $A \in \mathbb{B}(\mathbb{H})$ and $B \in \mathbb{B}(\mathbb{K})$ as a block-matrix form:

$$A = [A_{ij}]_{i,j=1}^{m,m} \quad \text{and} \quad B = [B_{kl}]_{k,l=1}^{n,n}$$

where $A_{ij} \in \mathbb{B}(\mathbb{H}_j, \mathbb{H}_i)$ and $B_{kl} \in \mathbb{B}(\mathbb{K}_l, \mathbb{K}_k)$ for each i, j, k, l.

2.1. Tracy-Singh Product for Operators

Let $A \in \mathbb{B}(\mathbb{H})$ and $B \in \mathbb{B}(\mathbb{K})$. Recall that the tensor product of A and B, denoted by $A \otimes B$, is a unique bounded linear operator on the tensor product space $\mathbb{H} \otimes \mathbb{K}$ such that

$$(A \otimes B)(x \otimes y) = Ax \otimes By, \quad \forall x \in \mathbb{H}, \forall y \in \mathbb{K}.$$

When $\mathbb{H} = \mathbb{K} = \mathbb{C}$, the tensor product of operators becomes the Kronecker product of matrices.

Definition 1. Let $A = [A_{ij}]_{i,j=1}^{m,m} \in \mathbb{B}(\mathbb{H})$ and $B = [B_{kl}]_{k,l=1}^{n,n} \in \mathbb{B}(\mathbb{K})$. The Tracy-Singh product of A and B is defined to be in the form

$$A \boxtimes B = \left[[A_{ij} \otimes B_{kl}]_{kl} \right]_{ij}, \tag{5}$$

which is a bounded linear operator from $\bigoplus_{i=1}^{m}\bigoplus_{k=1}^{n} \mathbb{H}_i \otimes \mathbb{K}_k$ into itself.

When $m = n = 1$, the Tracy-Singh product $A \boxtimes B$ is the tensor product $A \otimes B$. If $\mathbb{H}_i = \mathbb{K}_j = \mathbb{C}$ for all i, j, the above definition becomes the usual Tracy-Singh product for complex matrices.

Lemma 1 ([10,11]). *Let A, B, C, D be compatible operators. Then*

1. $(\alpha A) \boxtimes B = A \boxtimes (\alpha B) = \alpha (A \boxtimes B)$ for any $\alpha \in \mathbb{C}$.
2. $(A + B) \boxtimes (C + D) = A \boxtimes C + A \boxtimes D + B \boxtimes C + B \boxtimes D$.
3. $(A \boxtimes B)(C \boxtimes D) = (AC) \boxtimes (BD)$.
4. If A and B are Hermitian, then so is $A \boxtimes B$.
5. If A and B are positive and invertible, then $(A \boxtimes B)^\alpha = A^\alpha \boxtimes B^\alpha$ for any $\alpha \in \mathbb{R}$.
6. If $A \geqslant C \geqslant 0$ and $B \geqslant D \geqslant 0$, then $A \boxtimes B \geqslant C \boxtimes D \geqslant 0$.

2.2. Bochner Integration

Let Ω be a locally compact Hausdorff (LCH) space equipped with a finite Radon measure μ. A family $\mathcal{A} = (A_t)_{t \in \Omega}$ of operators in $\mathbb{B}(\mathbb{H})$ is said to be bounded if there is a constant $M > 0$ for which $\|A_t\| \leqslant M$ for all $t \in \Omega$. The family $\mathcal{A}$ is said to be a continuous field if parametrization $t \mapsto A_t$ is norm-continuous on Ω. Every continuous field $\mathcal{A} = (A_t)_{t\in\Omega}$ can have the Bochner integral $\int_\Omega A_t d\mu(t)$ if the norm function $t \mapsto \|A_t\|$ possess the Lebesgue integrability. In this case, the resulting integral is a unique element in $\mathbb{B}(\mathbb{H})$ such that

$$\phi\left(\int_\Omega A_t d\mu(t)\right) = \int_\Omega \phi(A_t) \, d\mu(t)$$

for every bounded linear functional ϕ on $\mathbb{B}(\mathbb{H})$.

Lemma 2 (e.g., [12]). *Let $(\mathbb{X}, \|\cdot\|_\mathbb{X})$ be a Banach space and (Γ, v) a finite measure space. Then a measurable function $f : \Gamma \to \mathbb{X}$ is Bochner integrable if and only if its norm function $\|f\|$ is Lebesgue integrable.*

Lemma 3 (e.g., [12]). *Let $f : \Gamma \to \mathbb{X}$ be a Bochner integrable function. If $\varphi : \mathbb{X} \to \mathbb{Y}$ is a bounded linear operator, then the composition $\varphi \circ f$ is Bochner integrable and*

$$\int_\Gamma (\varphi \circ f) dv = \varphi \int_\Gamma f dv.$$

Proposition 1. *Let $(A_t)_{t\in\Omega}$ be a bounded continuous field of operators in $\mathbb{B}(\mathbb{H})$. Then for any $X \in \mathbb{B}(\mathbb{K})$,*

$$\int_\Omega A_t d\mu(t) \boxtimes X = \int_\Omega (A_t \boxtimes X) d\mu(t).$$

Proof. Since the map $t \mapsto A_t$ is continuous and bounded, it is Bochner integrable on Ω. Note that the map $T \mapsto T \boxtimes X$ is linear and bounded by Lemma 1. Now, Lemma 3 implies that the map $t \mapsto A_t \boxtimes X$ is Bochner integrable on Ω and

$$\int_\Omega A_t d\mu(t) \boxtimes X = \int_\Omega (A_t \boxtimes X) d\mu(t).$$

for all $X \in \mathbb{B}(\mathbb{K})$. □

3. Chebyshev Type Inequalities Involving Tracy-Singh Products of Operators

From now on, let Ω be an LCH space equipped with a finite Radon measure μ. Let $\mathcal{A} = (A_t)_{t\in\Omega}$, $\mathcal{B} = (B_t)_{t\in\Omega}$, $\mathcal{C} = (C_t)_{t\in\Omega}$ and $\mathcal{D} = (D_t)_{t\in\Omega}$ be continuous fields of Hilbert space operators.

Definition 2. *The fields $\mathcal{A}$ and $\mathcal{B}$ are said to have the synchronous Tracy-Singh property if, for all $s, t \in \Omega$,*

$$(A_t - A_s) \boxtimes (B_t - B_s) \geq 0. \tag{6}$$

They are said to have the opposite-synchronous Tracy-Singh property if the reverse of (6) holds for all $s, t \in \Omega$.

Theorem 1. *Let $\mathcal{A}$ and $\mathcal{B}$ be bounded continuous fields of Hermitian operators in $\mathbb{B}(\mathbb{H})$ and $\mathbb{B}(\mathbb{K})$, respectively, and let $\alpha : \Omega \to [0, \infty)$ be a bounded measurable function.*

1. *If $\mathcal{A}$ and $\mathcal{B}$ have the synchronous Tracy-Singh property, then*

$$\int_\Omega \alpha(s) d\mu(s) \int_\Omega \alpha(t)(A_t \boxtimes B_t) d\mu(t) \geq \int_\Omega \alpha(t) A_t d\mu(t) \boxtimes \int_\Omega \alpha(s) B_s d\mu(s). \tag{7}$$

2. *If $\mathcal{A}$ and $\mathcal{B}$ have the opposite-synchronous Tracy-Singh property, then the reverse of (7) holds.*

Proof. By using Lemma 1, Proposition 1 and Fubini's Theorem [13], we have

$$\int_\Omega \alpha(s)d\mu(s) \int_\Omega \alpha(t)(A_t \boxtimes B_t)d\mu(t) - \int_\Omega \alpha(t)A_t d\mu(t) \boxtimes \int_\Omega \alpha(s)B_s d\mu(s)$$
$$= \iint_{\Omega^2} \alpha(s)\alpha(t)(A_t \boxtimes B_t)d\mu(t)d\mu(s) - \iint_{\Omega^2} \alpha(t)\alpha(s)(A_t \boxtimes B_s)d\mu(t)d\mu(s)$$
$$= \frac{1}{2}\iint_{\Omega^2} [\alpha(s)\alpha(t)(A_t \boxtimes B_t) - \alpha(t)\alpha(s)(A_t \boxtimes B_s)]\,d\mu(t)d\mu(s)$$
$$+ \frac{1}{2}\iint_{\Omega^2} [\alpha(t)\alpha(s)(A_s \boxtimes B_s) - \alpha(s)\alpha(t)(A_s \boxtimes B_t)]\,d\mu(s)d\mu(t)$$
$$= \frac{1}{2}\iint_{\Omega^2} \alpha(s)\alpha(t)\,[(A_t - A_s) \boxtimes (B_t - B_s)]\,d\mu(t)d\mu(s).$$

For the case 1, we have

$$\iint_{\Omega^2} \alpha(s)\alpha(t)\,[(A_t - A_s) \boxtimes (B_t - B_s)]\,d\mu(t)d\mu(s) \geq 0 \qquad (8)$$

and thus (7) holds. For another case, we get the reverse of (8) and, thus, the reverse of (7) holds. □

Remark 1. *In Theorem 1 and other results in this paper, we may assume that Ω is a compact Hausdorff space. In this case, every continuous field on Ω is automatically bounded.*

The next corollary is a discrete version of Theorem 1.

Corollary 1. *Let A_i, B_i be Hermitian operators and let ω_i be nonnegative numbers for each $i = 1, \ldots, n$. Let $\mathcal{A} = (A_1, \ldots, A_n)$ and $\mathcal{B} = (B_1, \ldots, B_n)$.*

1. *If $\mathcal{A}$ and $\mathcal{B}$ have the synchronous Tracy-Singh property, then*

$$\sum_{i=1}^n \omega_i \sum_{i=1}^n \omega_i(A_i \boxtimes B_i) \geq \left(\sum_{i=1}^n \omega_i A_i\right) \boxtimes \left(\sum_{i=1}^n \omega_i B_i\right). \qquad (9)$$

2. *If $\mathcal{A}$ and $\mathcal{B}$ have the opposite-synchronous Tracy-Singh property, then the reverse of (9) holds.*

Proof. From the previous theorem, set $\Omega = \{1, \ldots, n\}$ equipped with the counting measure and $\alpha(i) = \omega_i$ for all $i = 1, \ldots, n$. □

4. Chebyshev Integral Inequalities Concerning Weighted Pythagorean Means of Operators

Throughout this section, the space Ω is equipped with a total ordering $\preccurlyeq$.

Definition 3. *We say that a field $\mathcal{A}$ is increasing (decreasing, resp.) whenever $s \preccurlyeq t$ implies $A_s \leq A_t$ ($A_s \geq A_t$, resp.).*

Definition 4. *Two ordered pairs (X_1, X_2) and (Y_1, Y_2) of Hermitian operators are said to have the synchronous property if either*

$$X_i \leq Y_i \text{ for } i = 1, 2, \text{ or } X_i \geq Y_i \text{ for } i = 1, 2.$$

The pairs (X_1, X_2) and (Y_1, Y_2) are said to have the opposite-synchronous property if either

$$X_1 \leq Y_1 \text{ and } X_2 \geq Y_2, \text{ or } X_1 \geq Y_1 \text{ and } X_2 \leq Y_2.$$

Definition 5. Let $\mathcal{A}, \mathcal{B}, \mathcal{C}, \mathcal{D}$ be continuous fields of Hermitian operators. Two ordered pairs $(\mathcal{A}, \mathcal{B})$ and $(\mathcal{C}, \mathcal{D})$ are said to have the *synchronous monotone property* if (A_t, B_t) and (C_t, D_t) have the synchronous property for all $t \in \Omega$. They are said to have the *opposite-synchronous monotone property* if (A_t, B_t) and (C_t, D_t) have the opposite-synchronous property for all $t \in \Omega$.

Let us recall the notions of weighted classical Pythagorean means for operators. Indeed, they are generalizations of three famous symmetric operator means as follows. For any $w \in [0,1]$, the w-weighted arithmetic mean of $A, B \in \mathbb{B}(\mathbb{H})$ is defined by

$$A \nabla_w B = (1-w)A + wB.$$

The w-weighted geometric mean and w-weighted harmonic mean of $A, B \in \mathbb{B}(\mathbb{H})^{++}$ are defined respectively by

$$A \sharp_w B = A^{\frac{1}{2}}(A^{-\frac{1}{2}} B A^{-\frac{1}{2}})^w A^{\frac{1}{2}},$$

$$A !_w B = \left[(1-w)A^{-1} + wB^{-1}\right]^{-1}.$$

For any $A, B \in \mathbb{B}(\mathbb{H})^+$, we define the w-weighted geometric mean and w-weighted harmonic mean of A and B to be

$$A \sharp_w B = \lim_{\varepsilon \to 0^+} (A + \varepsilon I) \sharp_w (B + \varepsilon I).$$

$$A !_w B = \lim_{\varepsilon \to 0^+} (A + \varepsilon I) !_w (B + \varepsilon I),$$

respectively. Here, the limits are taken in the strong-operator topology.

Lemma 4 (see e.g., [14]). *The weighted geometric means, weighted arithmetic means and weighted harmonic means for operators are monotone in the sense that if $A_1 \leqslant A_2$ and $B_1 \leqslant B_2$, then $A_1 \sigma B_1 \leqslant A_2 \sigma B_2$ where σ is any of $\nabla_w, !_w, \sharp_w$.*

Lemma 5 ([15]). *Let $A, B, C, D \in \mathbb{B}(\mathbb{H})^+$ and $w \in [0, 1]$. Then*

$$(A \boxtimes B) \sharp_w (C \boxtimes D) = (A \sharp_w C) \boxtimes (B \sharp_w D).$$

Theorem 2. *Let $\mathcal{A}, \mathcal{B}, \mathcal{C}, \mathcal{D}$ be bounded continuous fields in $\mathbb{B}(\mathbb{H})^+$ and let $\alpha : \Omega \to [0, \infty)$ be a bounded measurable function.*

1. *If $\mathcal{A}, \mathcal{B}, \mathcal{C}, \mathcal{D}$ are either all increasing, or all decreasing then*

$$\int_\Omega \alpha(s) d\mu(s) \int_\Omega \alpha(t)[(A_t \boxtimes B_t) \sharp_w (C_t \boxtimes D_t)] d\mu(t)$$
$$\geqslant \int_\Omega \alpha(t)(A_t \sharp_w C_t) d\mu(t) \boxtimes \int_\Omega \alpha(s)(B_s \sharp_w D_s) d\mu(s). \quad (10)$$

2. *The reverse of (10) holds if either*

    2.1 *$\mathcal{A}, \mathcal{C}$ are increasing and $\mathcal{B}, \mathcal{D}$ are decreasing, or*
    2.2 *$\mathcal{A}, \mathcal{C}$ are decreasing and $\mathcal{B}, \mathcal{D}$ are increasing.*

Proof. Let $s, t \in \Omega$ and assume without loss of generally that $s \preccurlyeq t$. By applying Lemmas 1 and 5, Proposition 1, and Fubini's Theorem [13], we have

$$\int_\Omega \alpha(s) d\mu(s) \int_\Omega \alpha(t)[(A_t \boxtimes B_t) \sharp_w (C_t \boxtimes D_t)] d\mu(t) - \int_\Omega \alpha(t)(A_t \sharp_w C_t) d\mu(t) \boxtimes \int_\Omega \alpha(s)(B_s \sharp_w D_s) d\mu(s)$$
$$= \iint_{\Omega^2} \alpha(s)\alpha(t)[(A_t \boxtimes B_t) \sharp_w (C_t \boxtimes D_t)] d\mu(t) d\mu(s)$$
$$\quad - \iint_{\Omega^2} \alpha(t)\alpha(s)[(A_t \sharp_w C_t) \boxtimes (B_s \sharp_w D_s)] d\mu(t) d\mu(s)$$
$$= \iint_{\Omega^2} \alpha(s)\alpha(t)[(A_t \sharp_w C_t) \boxtimes (B_t \sharp_w D_t)] d\mu(t) d\mu(s)$$
$$\quad - \iint_{\Omega^2} \alpha(t)\alpha(s)[(A_t \sharp_w C_t) \boxtimes (B_s \sharp_w D_s)] d\mu(t) d\mu(s)$$
$$= \frac{1}{2} \iint_{\Omega^2} \alpha(s)\alpha(t)[(A_t \sharp_w C_t) \boxtimes (B_t \sharp_w D_t) - (A_t \sharp_w C_t) \boxtimes (B_s \sharp_w D_s)] d\mu(t) d\mu(s)$$
$$\quad + \frac{1}{2} \iint_{\Omega^2} \alpha(t)\alpha(s)[(A_s \sharp_w C_s) \boxtimes (B_s \sharp_w D_s) - (A_s \sharp_w C_s) \boxtimes (B_t \sharp_w D_t)] d\mu(s) d\mu(t)$$
$$= \frac{1}{2} \iint_{\Omega^2} \alpha(s)\alpha(t)[A_t \sharp_w C_t - A_s \sharp_w C_s] \boxtimes [B_t \sharp_w D_t - B_s \sharp_w D_s] d\mu(t) d\mu(s).$$

If A, B, C, D are all increasing, we have by Lemma 4 that $A_t \sharp_w C_t \geqslant A_s \sharp_w C_s$ and $B_t \sharp_w D_t \geqslant B_s \sharp_w D_s$. If A, B, C, D are all decreasing, we have $A_t \sharp_w C_t \leqslant A_s \sharp_w C_s$ and $B_t \sharp_w D_t \leqslant B_s \sharp_w D_s$. Both cases lead to the same conclusion that

$$(A_t \sharp_w C_t - A_s \sharp_w C_s) \boxtimes (B_t \sharp_w D_t - B_s \sharp_w D_s) \geqslant 0,$$

and hence (10) holds. The cases 2.1 and 2.2 yield the same conclusion that

$$(A_t \sharp_w C_t - A_s \sharp_w C_s) \boxtimes (B_t \sharp_w D_t - B_s \sharp_w D_s) \leqslant 0.$$

and hence the reverse of (10) holds. □

Lemma 6. *Let A, B, C, D be Hermitian operators in $\mathbb{B}(\mathbb{H})$ and $w \in [0, 1]$.*

1. *If (A, B) and (C, D) have the synchronous property, then*

$$(A \boxtimes B) \nabla_w (C \boxtimes D) \geqslant (A \nabla_w C) \boxtimes (B \nabla_w D). \quad (11)$$

2. *If (A, B) and (C, D) have the opposite-synchronous property, then the reverse of (11) holds.*

Proof. For the synchronous case, we have by using positivity of the Tracy-Singh product (Lemma 1) that $(A - C) \boxtimes (B - D) \geqslant 0$. Applying Lemma 1, we obtain

$$0 \leqslant w(1-w) \left[(A_1 - B_1) \boxtimes (A_2 - B_2) \right]$$
$$= w(1-w) \left[A_1 \boxtimes A_2 - A_1 \boxtimes B_2 - B_1 \boxtimes A_2 + B_1 \boxtimes B_2 \right]$$
$$= \left[(1-w)(A_1 \boxtimes A_2) + w(B_1 \boxtimes B_2) \right] - \left[(1-w)A_1 + wB_1 \right] \boxtimes \left[(1-w)A_2 + wB_2 \right]$$
$$= \left[(A_1 \boxtimes A_2) \nabla_w (B_1 \boxtimes B_2) \right] - \left[(A_1 \nabla_w B_1) \boxtimes (A_2 \nabla_w B_2) \right].$$

Thus $(A_1 \nabla_w B_1) \boxtimes (A_2 \nabla_w B_2) \leqslant (A_1 \boxtimes A_2) \nabla_w (B_1 \boxtimes B_2)$.

For the opposite-synchronous case, we have $(A_1 - B_1) \boxtimes (A_2 - B_2) \leq 0$ and hence the reverse of inequality (11) holds. □

Theorem 3. *Let $\mathcal{A}, \mathcal{B}, \mathcal{C}, \mathcal{D}$ be bounded continuous fields of operators in $\mathbb{B}(\mathbb{H})^+$, let $\alpha : \Omega \to [0, \infty)$ be a bounded measurable function.*

1. *If $(\mathcal{A}, \mathcal{B})$ and $(\mathcal{C}, \mathcal{D})$ have the synchronous monotone property and all of $\mathcal{A}, \mathcal{B}, \mathcal{C}, \mathcal{D}$ are either increasing or decreasing, then*

$$\int_\Omega \alpha(s) d\mu(s) \int_\Omega \alpha(t)[(A_t \boxtimes B_t) \nabla_w (C_t \boxtimes D_t)] d\mu(t) \geq \int_\Omega \alpha(t)(A_t \nabla_w C_t) d\mu(t) \boxtimes \int_\Omega \alpha(s)(B_s \nabla_w D_s) d\mu(s). \quad (12)$$

2. *If $(\mathcal{A}, \mathcal{B})$ and $(\mathcal{C}, \mathcal{D})$ have the opposite-synchronous monotone property and if either*

   2.1 *$\mathcal{A}, \mathcal{C}$ are increasing and $\mathcal{B}, \mathcal{D}$ are decreasing, or*
   2.2 *$\mathcal{A}, \mathcal{C}$ are decreasing and $\mathcal{B}, \mathcal{D}$ are increasing,*

   then the reverse of (12) holds.

Proof. Let $s, t \in \Omega$ and assume without loss of generally that $s \preccurlyeq t$. First, we consider the case 1. We have by using Lemmas 1 and 6, proposition 1, and Fubini's Theorem [13] that

$$\int_\Omega \alpha(s) d\mu(s) \int_\Omega \alpha(t)[(A_t \boxtimes B_t) \nabla_w (C_t \boxtimes D_t)] d\mu(t) - \int_\Omega \alpha(t)(A_t \nabla_w C_t) d\mu(t) \boxtimes \int_\Omega \alpha(s)(B_s \nabla_w D_s) d\mu(s)$$

$$= \iint_{\Omega^2} \alpha(s)\alpha(t)[(A_t \boxtimes B_t) \nabla_w (C_t \boxtimes D_t)] d\mu(t) d\mu(s)$$

$$- \iint_{\Omega^2} \alpha(t)\alpha(s)[(A_t \nabla_w C_t) \boxtimes (B_s \nabla_w D_s)] d\mu(t) d\mu(s)$$

$$\geq \iint_{\Omega^2} \alpha(s)\alpha(t)[(A_t \nabla_w C_t) \boxtimes (B_t \nabla_w D_t)] d\mu(t) d\mu(s)$$

$$- \iint_{\Omega^2} \alpha(t)\alpha(s)[(A_t \nabla_w C_t) \boxtimes (B_s \nabla_w D_s)] d\mu(t) d\mu(s)$$

$$= \iint_{\Omega^2} \alpha(s)\alpha(t)[(A_t \nabla_w C_t) \boxtimes (B_t \nabla_w D_t) - (A_t \nabla_w C_t) \boxtimes (B_s \nabla_w D_s)] d\mu(t) d\mu(s)$$

$$= \frac{1}{2} \iint_{\Omega^2} \alpha(s)\alpha(t)[(A_t \nabla_w C_t) - (A_s \nabla_w C_s)] \boxtimes [(B_t \nabla_w D_t) - (B_s \nabla_w D_s)] d\mu(t) d\mu(s).$$

Now, by Lemmas 1 and 4, we have

$$(A_t \nabla_w C_t - A_s \nabla_w C_s) \boxtimes (B_t \nabla_w D_t - B_s \nabla_w D_s) \geq 0$$

and hence (12) holds. The case 2 can be similarly proven. □

Lemma 7. *Let A, B, C, D be positive operators in $\mathbb{B}(\mathbb{H})$ and $w \in [0, 1]$.*

1. *If (A, B) and (C, D) are synchronous, then*

$$(A \boxtimes B) !_w (C \boxtimes D) \leq (A !_w C) \boxtimes (B !_w D). \quad (13)$$

2. *If (A, B) and (C, D) are opposite-synchronous, then the reverse of (13) holds.*

Proof. Assume that $(\mathcal{A}, \mathcal{B})$ and $(\mathcal{C}, \mathcal{D})$ are synchronous. By continuity, we may assume that $A, B, C, D > 0$. We have

$$(A^{-1} - C^{-1}) \boxtimes (B^{-1} - D^{-1}) \geq 0. \tag{14}$$

Using Lemma 1 and (14), we get

$$\begin{aligned}
0 &\leq w(1-w)A^{-1} \boxtimes B^{-1} + w(1-w)C^{-1} \boxtimes D^{-1} - w(1-w)A^{-1} \boxtimes D^{-1} - w(1-w)C^{-1} \boxtimes B^{-1} \\
&= \left[(1-w) - (1-w)^2\right] A^{-1} \boxtimes B^{-1} + (w - w^2)C^{-1} \boxtimes D^{-1} - w(1-w)A^{-1} \boxtimes D^{-1} \\
&\quad - w(1-w)C^{-1} \boxtimes B^{-1} \\
&= \left(A^{-1} \boxtimes B^{-1}\right) \nabla_w \left(C^{-1} \boxtimes D^{-1}\right) - \left(A^{-1} \nabla_w C^{-1}\right) \boxtimes \left(B^{-1} \nabla_w D^{-1}\right).
\end{aligned}$$

This implies that

$$\left(A^{-1} \boxtimes B^{-1}\right) \nabla_w \left(C^{-1} \boxtimes D^{-1}\right) \geq \left(A^{-1} \nabla_w C^{-1}\right) \boxtimes \left(B^{-1} \nabla_w D^{-1}\right).$$

Hence,

$$\begin{aligned}
(A \boxtimes B) !_w (C \boxtimes D) &= \left\{(A \boxtimes B)^{-1} \nabla_w (C \boxtimes D)^{-1}\right\}^{-1} \\
&= \left\{\left(A^{-1} \boxtimes B^{-1}\right) \nabla_w \left(C^{-1} \boxtimes D^{-1}\right)\right\}^{-1} \\
&\leq \left\{\left(A^{-1} \nabla_w C^{-1}\right) \boxtimes \left(B^{-1} \nabla_w D^{-1}\right)\right\}^{-1} \\
&= \left(A^{-1} \nabla_w C^{-1}\right)^{-1} \boxtimes \left(B^{-1} \nabla_w D^{-1}\right)^{-1} \\
&= (A !_w C) \boxtimes (B !_w D).
\end{aligned}$$

For the opposite-synchronous case, we have

$$(A^{-1} - C^{-1}) \boxtimes (B^{-1} - D^{-1}) \leq 0$$

and hence the reverse of (13) holds. □

Theorem 4. *Let $\mathcal{A}, \mathcal{B}, \mathcal{C}, \mathcal{D}$ be bounded continuous fields of operators in $\mathbb{B}(\mathbb{H})^+$ and $\alpha : \Omega \to [0, \infty)$ be a bounded measurable function.*

1. *If $(\mathcal{A}, \mathcal{B})$ and $(\mathcal{C}, \mathcal{D})$ have the opposite-synchronous monotone property and if all of $\mathcal{A}, \mathcal{B}, \mathcal{C}, \mathcal{D}$ are either increasing or decreasing, then*

$$\int_\Omega \alpha(s)d\mu(s) \int_\Omega \alpha(t)[(A_t \boxtimes B_t) !_w (C_t \boxtimes D_t)]d\mu(t) \\
\geq \int_\Omega \alpha(t)(A_t !_w C_t)d\mu(t) \boxtimes \int_\Omega \alpha(s)(B_s !_w D_s)d\mu(s). \tag{15}$$

2. *If $(\mathcal{A}, \mathcal{B})$ and $(\mathcal{C}, \mathcal{D})$ have synchronous monotone property and if either*

    2.1 $\mathcal{A}, \mathcal{C}$ *are both increasing, and $\mathcal{B}, \mathcal{D}$ are both decreasing, or*
    2.2 $\mathcal{A}, \mathcal{C}$ *are both decreasing and $\mathcal{B}, \mathcal{D}$ are both increasing,*

    then the reverse of (15) holds.

Proof. Let $s, t \in \Omega$ with $s \preccurlyeq t$. If the pairs $(\mathcal{A}, \mathcal{B})$ and $(\mathcal{C}, \mathcal{D})$ are opposite-synchronous, then we have by applying Lemmas 1 and 7, Proposition 1, and Fubini's Theorem [13] that

$$\int_\Omega \alpha(s) d\mu(s) \int_\Omega \alpha(t)[(A_t \boxtimes B_t) !_w (C_t \boxtimes D_t)] d\mu(t) - \int_\Omega \alpha(t)(A_t !_w C_t) d\mu(t) \boxtimes \int_\Omega \alpha(s)(B_s !_w D_s) d\mu(s)$$

$$= \iint_{\Omega^2} \alpha(s)\alpha(t)[(A_t \boxtimes B_t) !_w (C_t \boxtimes D_t)] d\mu(t) d\mu(s)$$

$$- \iint_{\Omega^2} \alpha(t)\alpha(s)[(A_t !_w C_t) \boxtimes (B_s !_w D_s)] d\mu(t) d\mu(s)$$

$$\geqslant \iint_{\Omega^2} \alpha(s)\alpha(t)[(A_t !_w C_t) \boxtimes (B_t !_w D_t)] d\mu(t) d\mu(s)$$

$$- \iint_{\Omega^2} \alpha(t)\alpha(s)[(A_t !_w C_t) \boxtimes (B_s !_w D_s)] d\mu(t) d\mu(s)$$

$$= \frac{1}{2} \iint_{\Omega^2} \alpha(s)\alpha(t)[A_t !_w C_t - A_s !_w C_s] \boxtimes [B_t !_w D_t - B_s !_w D_s] d\mu(t) d\mu(s).$$

For the case 1, we have, by Lemmas 1 and 4,

$$(A_t !_w C_t - A_s !_w C_s) \boxtimes (B_t !_w D_t - B_s !_w D_s) \geqslant 0$$

and hence (15) holds. Another assertion can be proved in a similar manner to that of the second assertion in Theorem 3. □

5. Chebyshev-Grüss Inequaities via Oscillations

Throughout this section, let Ω be an LCH space equipped with a probability Radon measure μ. For any continuous field $\mathcal{A} = (A_t)_{t \in \Omega}$ in $\mathbb{B}(\mathbb{H})$ and $\mathcal{B} = (B_t)_{t \in \Omega}$ in $\mathbb{B}(\mathbb{K})$, we define

$$\mathcal{A} \boxtimes \mathcal{B} = (A_t \boxtimes B_t)_{t \in \Omega}, \quad \mathcal{I}(\mathcal{A}) = \int_\Omega A_t d\mu(t),$$

$$\text{osc}(\mathcal{A}) = \max\{\|A_t - A_s\| : (t,s) \in \text{supp}(\mu \times \mu)\}.$$

Here, we recall that the support of the product measure $\mu \times \mu$ is defined by

$$\text{supp}(\mu \times \mu) = \{(t,s) \in \Omega^2 : (\mu \times \mu)(G) > 0 \text{ for all open sets } G \subseteq \Omega^2 \text{ containing } (t,s)\}.$$

We call $\text{osc}(\mathcal{A})$ the oscillation of the field $\mathcal{A}$.

Theorem 5. *Let $\mathcal{A} = (A_t)_{t \in \Omega}$ and $\mathcal{B} = (B_t)_{t \in \Omega}$ be continuous fields of Hermitian operators in $\mathbb{B}(\mathbb{H})$ and $\mathbb{B}(\mathbb{K})$, respectively. Then*

$$\mathcal{I}(\mathcal{A} \boxtimes \mathcal{B}) - \mathcal{I}(\mathcal{A}) \boxtimes \mathcal{I}(\mathcal{B}) \leqslant \frac{1}{2} \text{osc}(\mathcal{A}) \cdot \text{osc}(\mathcal{B})(\mu \times \mu)(\Omega^2 \setminus \Delta)(I_\mathbb{H} \boxtimes I_\mathbb{K}), \quad (16)$$

where $\Delta = \{(t,t) : t \in \Omega\}$.

Proof. We have by using Lemma 1, Proposition 1 and Fubini's Theorem [13] that

$$\begin{aligned}
\mathcal{I}(\mathcal{A} \boxtimes \mathcal{B}) - \mathcal{I}(\mathcal{A}) \boxtimes \mathcal{I}(\mathcal{B}) &= \int_\Omega d\mu(s) \int_\Omega A_t \boxtimes B_t d\mu(t) - \int_\Omega A_t d\mu(t) \boxtimes \int_\Omega B_s d\mu(s) \\
&= \iint_{\Omega^2} A_t \boxtimes B_t d\mu(t) d\mu(s) - \iint_{\Omega^2} A_t \boxtimes B_s d\mu(t) d\mu(s) \\
&= \frac{1}{2} \iint_{\Omega^2} (A_t \boxtimes B_t - A_t \boxtimes B_s + A_s \boxtimes B_s - A_s \boxtimes B_t) d\mu(t) d\mu(s) \\
&= \frac{1}{2} \iint_{\Omega^2 \setminus \Delta} (A_t - A_s) \boxtimes (B_t - B_s) d\mu(t) d\mu(s) \\
&\leq \frac{1}{2} \operatorname{osc}(\mathcal{A}) \cdot \operatorname{osc}(\mathcal{B}) (\mu \times \mu)(\Omega^2 \setminus \Delta)(I_\mathbb{H} \boxtimes I_\mathbb{K}). \quad \square
\end{aligned}$$

Corollary 2. Let $A_i \in \mathbb{B}(\mathbb{H})$ and $B_i \in \mathbb{B}(\mathbb{K})$ be Hermitian operators for all $i = 1, \ldots, n$. Then

$$\sum_{i=1}^n (A_i \boxtimes B_i) - \left(\sum_{i=1}^n A_i\right) \boxtimes \left(\sum_{i=1}^n B_i\right) \leq \frac{n(n-1)}{2} \max_{1 \leq i,j \leq n} \|A_i - A_j\| \cdot \max_{1 \leq i,j \leq n} \|B_i - B_j\|(I_\mathbb{H} \boxtimes I_\mathbb{K}).$$

Proof. Set $\Omega = \{1, \ldots, n\}$ equipped with the counting measure. We have

$$(\mu \times \mu)(\Omega^2 \setminus \Delta) = \frac{n(n-1)}{2}, \quad \operatorname{supp}(\mu \times \mu) = \Omega \times \Omega$$

and thus

$$\operatorname{osc}(A_1, \ldots, A_n) = \max_{1 \leq i,j \leq n} \|A_i - A_j\|, \quad \operatorname{osc}(B_1, \ldots, B_n) = \max_{1 \leq i,j \leq n} \|B_i - B_j\|. \quad \square$$

Example 1. Let $\Omega = [0,1]$, $w \in \Omega$ and $0 < \alpha \leq 1$. Consider the probability Radon measure $\mu = \alpha \lambda + (1 - \alpha)\delta_w$, where λ is Lebesgue measure on Ω and δ_w is the Dirac measure at w. Set

$$\mathcal{I}(\mathcal{A}) := \int_0^1 A_t d\mu(t) = \alpha \int_0^1 A_t d\lambda(t) + (1 - \alpha)A_w.$$

We have

$$\mu \times \mu = \alpha^2(\lambda \times \lambda) + \alpha(1 - \alpha)(\lambda \times \delta_w) + (1 - \alpha)\alpha(\delta_w \times \lambda) + (1 - \alpha)^2(\delta_w \times \delta_w).$$

Then $\operatorname{supp}(\mu \times \mu) = [0,1] \times [0,1]$ and $(\mu \times \mu)([0,1]^2 \setminus \Delta) = \alpha(2 - \alpha)$. For any continuous fields $\mathcal{A} = (A_t)_{t \in \Omega}$ and $\mathcal{B} = (B_t)_{t \in \Omega}$ of Hermitian operators, the inequality (16) becomes

$$\mathcal{I}(\mathcal{A} \boxtimes \mathcal{B}) - \mathcal{I}(\mathcal{A}) \boxtimes \mathcal{I}(\mathcal{B}) \leq \frac{1}{2}\alpha(2 - \alpha) \max_{0 \leq s,t \leq 1} \|A_t - A_s\| \cdot \max_{0 \leq t,s \leq 1} \|B_t - B_s\|(I_\mathbb{H} \boxtimes I_\mathbb{K}).$$

6. Conclusions

We establish several integral inequalities of Chebyshev type for continuous fields of Hermitian operators which are parametrized by an LCH space equipped with a finite Radon measure. We also obtain the Chebyshev-Grüss integral inequality via oscillations with respect to a probability Radon measure. These inequalities involve Tracy-Singh products and weighted versions of famous symmetric means. For a particular case that the LCH space is a finite space equipped with the counting measure, such integral inequalities reduce to discrete inequalities. Our results include Chebyshev-type inequalities for tensor product of operators and Tracy-Singh/Kronecker products of matrices.

Author Contributions: All authors contributed equally and significantly in writing this article. All authors read and approved the final manuscript.

Funding: The first author expresses his gratitude towards Thailand Research Fund for providing the Royal Golden Jubilee Ph.D. Scholarship, grant no. PHD60K0225 to support his Ph.D. study.

Acknowledgments: This research was supported by King Mongkut's Institute of Technology Ladkrabang.

Conflicts of Interest: The authors declare no conflict of interest.

References

1. Matharu, J.S.; Aujla, J.S. Hadamard product versions of the Chebyshev and Kantorovich inequalities. *J. Inequal. Pure Appl. Math.* **2009**, *10*, 6.
2. Mitrinović, D.S.; Pečarić, J.E.; Fink, A.M. *Classical and New Inequalities in Analysis*; Kluwer Academic: Dordrecht, The Netherlands, 1993.
3. Moslehiana, M.S.; Bakherad, M. Chebyshev type inequalities for Hilbert space operators. *J. Math. Anal. Appl.* **2014**, *420*, 737–749. [CrossRef]
4. Grüss, G. Über das Maximum des absoluten Betrages von $\frac{1}{b-a}\int_a^b f(x)g(x)dx - \frac{1}{(b-a)^2}\int_a^b f(x)dx \cdot \int_a^b g(x)dx$. *Mathematische Zeitschrift* **1935**, *39*, 215–226. [CrossRef]
5. Gavrea, B. Improvement of some inequalities of Chebysev-Grüss type. *Comput. Math. Appl.* **2012**, *64*, 2003–2010. [CrossRef]
6. Acu, A.M.; Rusu, M.D. New results concerning Chebyshev–Grüss-type inequalities via discrete oscillations. *Appl. Math. Comput.* **2014**, *243*, 585–593. [CrossRef]
7. Gonska, H.; Raçsa, I.; Rusu, M.D. Chebyshev–Grüss-type inequalities via discrete oscillations. *Bul. Acad. Stiinte Repub. Mold. Mat.* **2014**, *1*, 63–89.
8. Kubrusly, C.S.; Vieira, P.C.M. Convergence and decomposition for tensor products of Hilbert space operators. *Oper. Matrices* **2008**, *2*, 407–416. [CrossRef]
9. Zanni, J.; Kubrsly, C.S. A note on compactness of tensor products. *Acta Math. Univ. Comenian. (N.S.)* **2015**, *84*, 59–62.
10. Ploymukda, A.; Chansangiam, P.; Lewkeeratiyutkul, W. Algebraic and order properties of Tracy–Singh products for operator matrices. *J. Comput. Anal. Appl.* **2018**, *24*, 656–664.
11. Ploymukda, A.; Chansangiam, P.; Lewkeeratiyutkul, W. Analytic properties of Tracy–Singh products for operator matrices. *J. Comput. Anal. Appl.* **2018**, *24*, 665–674.
12. Aliprantis, C.D.; Border, K.C. *Infinite Dimensional Analysis: A Hitchhiker's Guide*; Springer-Verlag: New York, NY, USA, 2006.
13. Bogdanowicz, W.M. Fubini theorems for generalized Lebesgue-Bochner-Stieltjes integral. *Proc. Jpn. Acad.* **1966**, *41*, 979–983. [CrossRef]
14. Kubo, F.; Ando, T. Means of positive linear operators. *Math. Ann.* **1980**, *246*, 205–224. [CrossRef]
15. Ploymukda, A.; Chansangiam, P. Geometric means and Tracy–Singh products for positive operators. *Commun. Math. Appl.* **2018**, *9*, 475–488.

© 2019 by the authors. Licensee MDPI, Basel, Switzerland. This article is an open access article distributed under the terms and conditions of the Creative Commons Attribution (CC BY) license (http://creativecommons.org/licenses/by/4.0/).

Article

Framework for Onboard Bus Comfort Level Predictions Using the Markov Chain Concept

Paweł Więcek, Daniel Kubek *, Jan Hipolit Aleksandrowicz and Aleksandra Stróżek

Department of Transportation Systems, Faculty of Civil Engineering, Cracow University of Technology, 31-155 Kraków, Poland; pwiecek@pk.edu.pl (P.W.); jaleksandrowicz@pk.edu.pl (J.H.A.); aleksandra.strozek@pk.edu.pl (A.S.)
* Correspondence: dkubek@pk.edu.pl; Tel.: +48-12-628-37-21

Received: 15 May 2019; Accepted: 31 May 2019; Published: 4 June 2019

Abstract: Efficiently functioning public transport has a significant positive impact on the entire transportation system performance through numerous aspects, such as the reduction of congestion, energy consumption, and emissions. In most cases, the basic elements of public transport are the bus transport subsystem. Currently, in addition to criteria such as punctuality, the frequency of departures, and the number of transfers, a travelling comfort level is an important element for passengers. An overcrowded bus may discourage travelers from choosing this mode of transport and induce them to use a private car despite the existence of many other facilities offered by a given public transport system. Therefore, the forecasting of bus passenger demand, as well as bus occupancy at individual bus stops, is currently an important research direction. The main goal of the article is to present the conceptual framework for the Advanced Travel Information System with the prediction module. The proposed approach assumes that the prediction module is based on the use of the Markov Chain concept. The efficiency and accuracy of the obtained prediction were presented based on a real-life example, where the measurements of passengers boarding and alighting at bus stops were made in a selected Cracow bus line. The methodology presented in the paper and the obtained results can significantly contribute to the development of solutions and systems for a better management as well as a cost and energy consumption optimisation in the public transport system. Current and forecasted information related to bus occupancy, when properly used in the travel information system, may have a positive impact on the development of urban mobility patterns by encouraging the use of public transport. This article addresses the current and practical research problem using an adequate theoretical mathematical tool to describe it, reflecting the characteristics and nature of the phenomenon being studied. To the best of the authors' knowledge, the article deals for the first time with the problem of prediction of onboard bus comfort levels based on in-vehicle occupancy.

Keywords: onboard comfort level; Markow chain; bus passenger occupancy prediction

1. Introduction

Efficiently functioning public transport has a significant positive impact on the entire transportation system performance through numerous aspects, such as the reduction of congestion, energy consumption, and emissions. In most cases, the fundamental elements of public transport are the bus transport subsystem. Currently, in addition to criteria such as punctuality, the frequency of departures, and the number of transfers, the comfort of travelling is an essential element for travelers. It may be significantly related to the degree of occupancy and capacity of the bus. An overcrowded bus may discourage travelers from choosing this mode of transport and induce him to use a private car despite the existence of many other facilities offered by a given public transport system. Therefore, the forecasting of bus passenger demand and the forecasting of bus occupancy at individual bus stops is currently an important research direction. The variability and cyclicity in passenger flows make such forecasts extremely useful. From the perspective of transport system planners, they can be used for an optimal allocation of resources and bus types in relation to current demand. From the traveler's point of view, the use of this type of forecast can help to reduce waiting times at the bus stop as well as helping him choose the right departure time. The forecasting methods and techniques used in this area mainly concern long-term prediction. Given the above, the main goal of the article is to present the framework of the prediction subsystem for the Advanced Travel Information System (ATIS), which enables one to predict the onboard comfort level related to the actual bus occupation.

The article is organized as follows. The first section is a literature review. It describes the conceptual framework for the ATIS subsystem with the comfort level prediction module. It also discusses the potential application of the proposed approach. The main core of the article concerns the forecasts in the public transportation system; hence, the second part of the literature review deals with that area. Subsequently, the theoretical background of the Markov chain (MC) for the prediction model is presented, defining the comfort level states. The characteristics and benefits of applying the MC for an onboard bus comfort level prediction based on a real-life case study summarise the previous theoretical discussion.

2. Literature Review

2.1. Predictive Framework for ATIS Subsystem

The principle of information is defined as one of the foundations of the city's transport policy. The Advanced Traveler Information System (ATIS) is a way to implement the information principle and is an integral part of the Intelligent Transport Systems (ITS) [1,2]. ATIS systems can use all transport data, including the traffic volume, journey times, and restrictions on selected sections, timetables, vehicles locations in the network, interchanges, traffic events, and weather conditions. The information can come both from the vehicles and traffic management centres [3,4].

Access to information from ATIS systems can be public or limited. Restrictions may result from the payment of access to the system or from the fact that the system is dedicated to selected users [5]. Research [6] shows that mobile phone users with Internet access are most likely to use information from ATIS systems, and the interest of travelers in accessing travel information is directly related to the usefulness of the presented data by ATIS systems [4].

The considered issue in the paper is one of the crucial problems, mainly when bus congestion results in the resignation of some passengers from travelling with public transport. Well-prepared forecasts make it possible to make good use of rolling stock, to plan a journey, and to manage urban public transport. Figure 1 presents a diagram of the information flow and relevance of a forecasting module within ATIS. The data based on which the model was created and tested came from buses operating in Cracow.

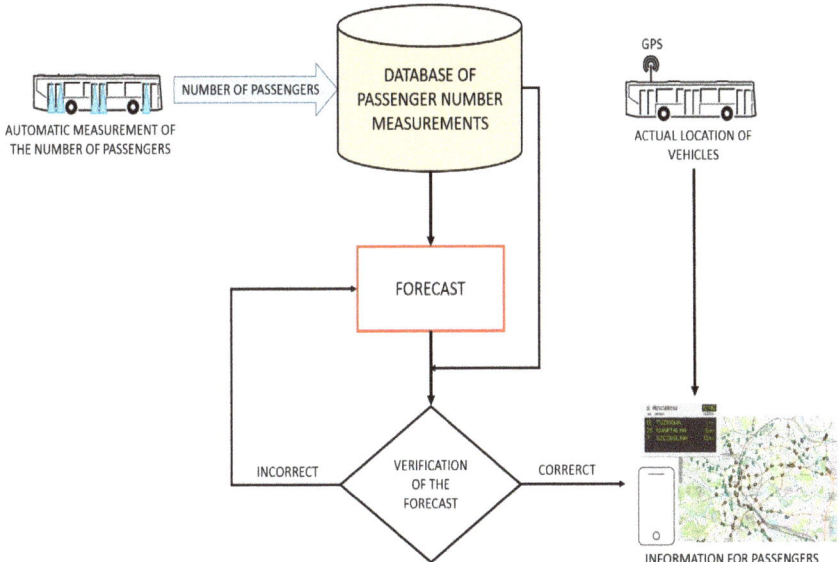

Figure 1. Diagram of the created Advanced Traveler Information System system.

The development of information technology has a significant impact on the functioning of cities and urban public transport. This has enabled the development in a short time of many mobile applications aimed at facilitating travel. They relate to trip planning, checking timetables, or information about the location of the exact vehicle based on the GPS signal. Common access to such extensive information has led to an increase in passenger requirements.

Moreover, information about the current or predicted passenger demand in public transport vehicles is becoming critical. The automatic passenger counting systems using buses do not provide the utilization of data in real time. However, the data will be sufficiently accurate to be used to produce a forecast. Such predictions will enable the optimization of the allocation of buses to public transport lines and the creation of applications for passengers (in the framework of ATIS).

The problem connected with bus allocations to public transport lines is important for carriers. The increasing number of rolling stocks types makes it more challenging to obtain an optimal solution. A large number of urban bus manufacturers and the choice of the cheapest offers by carriers make it difficult to maintain a uniform fleet of vehicles. Vehicles of the same dimensions may differ in travel comfort with the same number of passengers. That is why, in order to achieve a high level of comfort for passengers with a maximum use of rolling stock, it is necessary to optimise the allocation of buses to public transport lines. The solutions that emerged within this topic are described in articles, the most important of which are presented in the table below (Table 1).

Table 1. Articles on the optimization of the allocation of rolling stock to public transport lines.

Issues Raised in the Papers	Articles
Optimal queuing of rolling stock at depots for line departure	Blasum M. Bussieck M.R., Hochstattler W., Moll C.H., Scheel H., Winter T. [7]
Optimising the number of vehicles serving the urban public transport system	Haase K., Deaulniers G., Denosiers J. [8] Kidwai F.A., Marwah B.R., Deb K., Karim M.R. [9]
Optimisation of the allocation of rolling stock to lines - environmental criteria	Jimenez F., Roman A., (2016), Li J.Q., Head K.L. [10] Li L., Lo H.K., Cen X. [11] Beltran B., Carrese S., Cipriani E., Petrelli M. [12]
Allocation of rolling stock to lines, as part of public transport planning, day by day planning	Lusby R.M., Larsen J., Bull S. [13]
Characteristics of management systems for the allocation of rolling stock to lines	Gancarz T. (1998), Papierkowski K. [14] Cejrowski M., Krych A., Pawłowski M. [15] Moreira J.M., de Sousa J.F. [16]
Optimisation of the allocation of rolling stock to lines to minimise fuel consumption	Oziomek J., Rogowski A. [17,18]

The optimisation of bus allocations to urban public transport lines is presented in different ways in scientific publications. First, the allocation of rolling stock to public transport lines is taken as the queuing of vehicles at depots [7]. However, this solution does not take into account the possibility of a rolling stock rotation between different depots. When there is more than one depot, it is impossible to obtain an optimal solution. Other articles define the allocation of rolling stock as scheduling that reduces costs and the number of vehicles that are needed [8,17,18]. This approach is characterised by the use of various optimisation methods, both traditional linear programming methods and heuristics [9]. However, it should be noted that the analysed models do not use data on the forecast of demand for transport services. In [13], the problem of a model that is resistant to fluctuations in parameters was noted (the issue of planning the allocation of rolling stock for each day). The solutions presented in the publication are based on the example of railway transport. Another type of criterion which is becoming more and more important due to the growing social awareness of environmental protection is the optimisation of the allocation of rolling stock as a means of minimising the environmental impact. This topic is described extensively in [10–12,19]. Finally, there are publications that describe the existing tools to support decision-making when scheduling the allocation of rolling stock to public transport lines [10,14,16,20].

2.2. Prediction Methods in Public Transportation

The main problem in public transportation, which researchers are trying to model and predict, is passenger demand. It has a direct influence on the efficiency of the public transport system, raising the competitiveness of this mean of transport and, lastly, fulfilling the expectations of passengers and encouraging them to choose this type of transport.

The problem of bus occupancy level forecasting is quite an important aspect, particularly for decision-makers. This is why there is an abundance of articles addressing this issue. A plurality of methods whose aim is to deal with this kind of forecasting shows an interest attached to the seriousness of this issue. Table 2 shows selected methodologies, types, modes of transport and the studies in which they were presented.

Table 2. Passenger demand and flow prediction studies in public transport.

Author(s)	Methodology	Type	Modes
Y. Mo, Y. Su [21]	Neural networks	Transit passenger Flow	Bus
Y. Li [22]	Grey Markov Chain model	Flow	Railway
S. Z. Zhao, T. H. Ni, Y. Wang, X. T. Gao [23]	Wavelet analysis, Neural networks	Flow	Transit system
L. Liu, R. C. Chen [24]	Deep learning method	Flow	Bus rapid transit
Y. Li, X. Wang, S. Sun, X. Ma, G. Lu [25]	Multiscale radial basis function networks	Flow	Subway
J. Zhang, D. Shen, L. Tu, F. Zhang, C. Xu, Y. Wang, C. Tian, X. Li, B. Huang, Z. Li [26]	Extended Kalman filter model	Flow	Bus transit system
Q. Chen, W. Li, J. Zhao [27]	Least Squares Support Vector Machine	Flow	Bus
R. Xue, D. J. Sun, S. Chen [28]	Time series and interactive multiple model (IMM)	Demand	Bus
C. Zhou, P. Dai, R. Li [29]	Time-varying Poisson model, Weighted time-varying Poisson model, ARIMA	Demand	Bus
Z. Ma, J. Xing, M. Mesbah, L. Ferreira [30]	Interactive Multiple, Model-based Pattern Hybrid (IMMPH)	Demand	Bus
T. H. Tsai, C. K. Lee, C. H. Wei [31]	Neural network	Demand	Railway
Z. Wang, C. Yang, C. Zang [32]	Hybrid model (BP neural network & time series model)	Flow prediction	Bus stop
J. Roos, S. Bonnevay, G. Gavin [33]	Dynamic Bayesian network	Flow forecasting	Metro
Z. Wei, Z. Jinfu [34]	Grey-Markov Method	Passenger traffic	Passenger turnover
Z. S. Xiao, B. H. Mao, T. Zhang [35]	Hybrid model—BP neural network and Markov Chain	Daily passenger volume	Rail transit station

According to Table 2, it is evident that there were not so many papers that tried to solve the passenger prediction problem using the Markov Chain method. Most of the presented methods are used to predict the passenger flow or demand in the short-term. The Markov chain method appears mostly in a combination of Markov and Grey models to forecast the passenger flow or as a hybrid model—the Back Propagation neural network and Markov model—to forecast the daily passenger volume in the rail transit station. This shows the existence of different fields of research on the most accurate methods for forecasting the passenger demand or flow to improve public transportation efficiency directly.

There are certain areas in public transportation in which prediction can be beneficial during not only the organisation process but also the adaptation to real conditions. These may include, besides the passenger demand or flow prediction, the bus arrival time [36] prediction. From the public transport organiser's point of view, the passenger structure prediction in the transportation corridor may also be useful. For this purpose, D. Wang, X. Sun and Y. Li utilized the Markov process [37]. In order to meet the expectations of public transport users, an important area of prediction is public transport trip flows [38].

3. Onboard Bus Comfort Level and Markov Chain Concept

3.1. Bus Comfort Level

Onboard bus comfort is an essential aspect of the satisfaction perceived by bus passengers. The quality of the bus transit in terms of passenger comfort is usually an extensive set of partial influence factors, which are very difficult to quantify. Hence, this is usually the reason why they are not included in the various indicators for assessing the quality of public transport. Therefore, the group of indicators describing the comfort of travel include the inconvenience of travel resulting from the limited availability of seats or even standing in the vehicle. The literature most frequently mentions the nuisance ratio, and the nuisance ratio or seat occupancy rate, for different reference levels in the form of standards for the number of seats in a vehicle. The driving discomfort coefficient μ_j determines how many times a journey by public transport in specific conditions is more onerous in comparison with a journey where the passenger sits and the filling of standing places is small (0.5 passenger/m²). The driving discomfort μ_j is calculated from the following formula (based on [39]):

$$\mu_j = 0.8 + 3.6 \times (q - 0.15)^2 \tag{1}$$

where q is the relative onboard occupation calculated as:

$$q = \frac{N}{C_N} \tag{2}$$

where N is the absolute onboard occupation (number of passengers in the vehicle), and C_N is the nominal capacity of the vehicle (using the area of standing places as 0.15 m²/person).

The onboard comfort level can be defined based on the calculated value of the driving discomfort μ_j, and it could be one of the below possibilities [39]:

1. Comfort level A (corresponding to a factor of discomfort $\mu_j < 0.8$)—means that: approximately 10–70% of the vehicle seats are occupied; each passenger has a guaranteed seating position without being forced to travel in the immediate vicinity of another passenger; passengers travel without difficulty in carrying luggage, trolleys, bicycles, etc.
2. Comfort level B (corresponding to a factor of discomfort $\mu_j \in [0.8, 1.0)$ means that: all or almost all seating positions are occupied (70–100%); possibility to easily carry a baggage, trolleys, bicycles, etc.
3. Comfort level C (corresponding to a factor of discomfort $\mu_j \in [1.0, 1.4)$ means that: the small number of standing places is occupied, but it is possible to have free movement within the vehicle: easy access to the punch (up to 2 persons/m²).
4. Comfort level D (corresponding to a factor of discomfort $\mu_j \in [1.4, 2.1)$ indicates that the onboard occupancy level results in a difficulty of free movement in the vehicle and in access problems to the punch (up to 4 persons/m²).
5. Comfort level E (corresponding to the discomfort factor $\mu_j \in [2.1, 3.4)$ indicates an already high onboard congestion causing very difficult access to the punch (up to 6–7 persons/m²).
6. Comfort level F (corresponding to a factor of discomfort $\mu_j \geq 3.4$ is characterised by: very high in-vehicle congestion, during which it is not possible to cancel the ticket; the ride involves a large physical effort, with standing passengers pressing into the seating area; there are large difficulties in closing the door and incidental damages to the closing device; it is necessary to give way to passengers getting off their seats (over 7 persons/m²).

3.2. Short-Term In-Vehicle Occupation Predictions Based on Markov Chains Model

The problem of in-vehicle occupancy forecasting in public transport may be closely related to previously defined comfort levels. The specificity of the inflow of passengers to the given bus line stops in the following hours is stochastic. Therefore, it seems appropriate to use discrete Markov processes to determine the expected occupancy level of the vehicle at subsequent departures from a given stop on the line. Markov's process is a sequence of random variables, in which the probability of what will happen depends only on the present state. In the considered issue, only Markov processes defined on a discrete space of states will be used (Markov chains).

Let us denote by $X = (X_0, X_1, \ldots)$ a sequence of discrete random variables. The value of the variable X_t will be called the state of the chain at the moment t. It is assumed that the set of states S is calculable. The finite set of states can be defined as the state space S as follows:

$$s \in S, \ S = \{s_1, s_2, \ldots, s_{k-1}, s_k\}, \ k < \infty \qquad (3)$$

The discrete timestamps used in the considered problem can be defined as follows:

$$t \in T, \ T = \{1, 2, \ldots, t_{max}\}, \ t_{max} \leq \infty \qquad (4)$$

Definition 1. *A sequence of random variables X is a Markov chain if the Markov condition is fulfilled:*

$$\mathbb{P}(X_t = s | X_0 = x_0, \ldots X_{t-1} = x_{t-1}) = \mathbb{P}(X_t = s | X_{t-1} = x_{t-1}) \wedge t \in T \wedge x_0, x_1, \ldots x_{t-1} \in S \qquad (5)$$

Thus, for the Markov chain, the distribution of the conditional probability of the position in the time step t depends only on the conditional probability of the position in the previous step and not on the previous trajectory points (history).

Definition 2. *Let $\mathcal{P}$ be a matrix of dimensions $(k \times k)$ and elements $\{p_{ij} : i, j = 1, \ldots k\}$. A sequence of random variables $(X_0, X_1, \ldots)$ with values from a finite set of states $S = \{s_1, s_2, \ldots, s_{k-1}, S_k\}$ is called the Markov process, with the transition matrix $\mathcal{P}$, if for each t, any $i, j \in \{1, \ldots k\}$ and all $i_0, \ldots i_{t-1} \in \{1, \ldots k\}$,*

$$\mathbb{P}(X_{t+1} = s_j | X_0 = s_{i_0}, X_1 = s_{i_1}, \ldots, X_{t-1} = s_{i_{t-1}}, X_t = s_i) = \mathbb{P}(X_{t+1} = s_j | X_t = s_i) = p_{ij}$$

The elements of the transition matrix p_{ij} fulfill the following conditions:

$$\wedge \ t \in T \ p_{ij} = \mathbb{P}(X_{t+1} = s_j | X_t = s_i)$$

$$p_{ij} \geq 0 \ \wedge \ i, j \in \{1, \ldots k\}$$

$$\wedge_i \sum_j p_{ij} = 1$$

Definition 3. *The Markov chain is homogenous when for each time stamp it is described by the same transition matrix $\mathcal{P}$. The transition matrix is fixed and does not depend on time.*

In the use of Markov chains, the initial state plays a crucial role. Formally, the initial state is a random variable X_0. Therefore, the Markov chain often starts with a certain probability distribution across the state space.

Definition 4. *The initial distribution is a vector defined as follows:*

$$D^{(0)} = \left[d_1^{(0)}, d_2^{(0)}, \ldots, d_k^{(0)}\right] = [\mathbb{P}(X_0 = s_1), \mathbb{P}(X_0 = s_2), \ldots, \mathbb{P}(X_0 = s_k)]$$

To determine the distribution of the forecasted state of the modelled object for the n-th time step ahead, the following equation can be used:

$$D^{(t+n)} = D^{(t)} \cdot \mathcal{P}^n$$

where n is the parameter defining the forecasting horizon.

4. Case Study

In order to verify the approach proposed in the article, a simulation experiment was carried out. The basis of the example was to estimate the forecasted state of the vehicle occupancy for a selected communication line with a given number of stops. In the example, a real data set from the automatic counting systems of the vehicle was used. The analysed time horizon covered two weeks for one of the most crowded bus lines in Cracow. The line under consideration belongs to one of the highest frequency levels and contain 19 bus stops. On business days, the number of trips on the line under consideration was $t_{max} = 68$, whereas on weekends $t_{max} = 47$. In the computational example, the forecasts of the occupation state $S = \{1, 2, \ldots 6\}$ were determined sequentially for one time step ahead (each single departure from the bus stop was a correspondingly successive time step). The state $s_1 = 1$ corresponds to the lowest level of vehicle occupancy, while $s_6 = 6$ denotes the highest. For each time step t, the initial state distribution $D^{(t)}$ has been updated on the basis of the available historical data. The elements of the transition matrix $\mathcal{P}$ were estimated empirically based on the historical data set individually for each bus stop in order to map its specificity and dynamics. The forecasted state was assumed to be the one for which the probability of occurrence in the forecasted state distribution $D^{(t+1)}$ was the highest. Figure 2 shows an exemplary adjustment of state forecasts to the real observed states of vehicle occupancy for a selected bus stop on a given day.

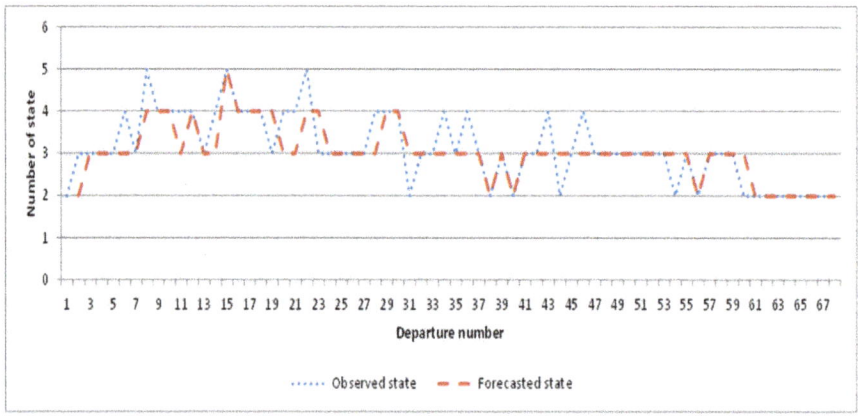

Figure 2. Adjustment of the forecasted vehicle occupancy states to the observed values for a given bus stop.

The presented sequence of observed vehicle occupancy states and received forecasts concerns the bus stop located in the second part of the analysed transport line. This is evidenced by the high variability of the observed states during the working day. The obtained forecast values, despite errors, try to keep up with the pace of changes in the observed time series.

The distribution of the root means square errors for each bus stop for the considered period is shown in Figure 3.

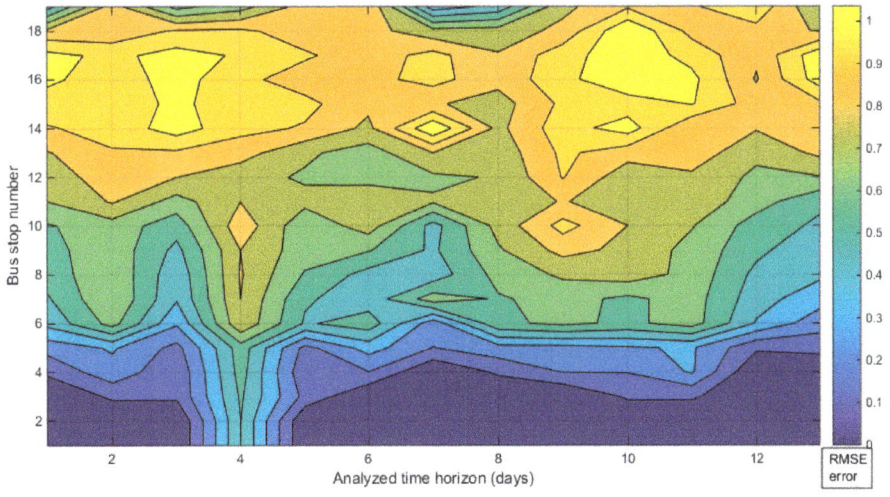

Figure 3. Distribution of root mean square errors for the analysed period.

The distribution of Root Mean Square Error (RMSE) errors received along the time horizon and bus stop number indicates that the highest values occur at the bus stops in the second part of the line journey (counting from the first stop). This results in the specificity of the analysed line, which passes through crucial areas in the city and numerous interchange nodes. This generates a greater randomness and variability among the incoming passengers, which leads to more significant forecasting errors. Lower errors characterise periods (t = 6 Saturday, t = 7 Sunday) due to the reduced number of trips.

In order to determine how often the model made an error and how much the predicted occupancy state of the vehicle differed from the observed state, an error histogram was prepared, as shown in Figure 4.

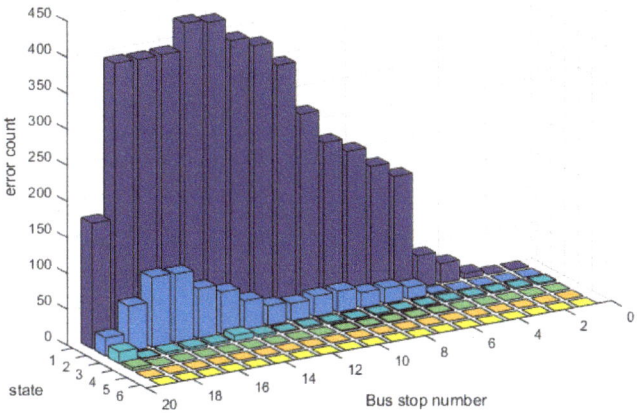

Figure 4. Prediction error histogram.

The histogram shows the frequency occurrence of a forecast error equal to $e_1 = 1$, $e_2 = 2$, ...$e_5 = 5$, where e_1 is the difference by one state, e_2 by two states, etc. In the period of time covered by the analysis, the most numerous group is the e_1 error set, where the obtained forecast differs only by one state from the observed real value. The second, much less numerous group is e_2. The sets of errors e_3 and e_4 constitute a small percentage of the whole population, while the remaining errors did not occur at all during the examined time horizon.

The averaged absolute percentage forecast errors for the relevant period and subsequent stops are presented in Table 3.

Table 3. Mean absolute percentage forecast errors.

		MEAN ABSOLUTE PERCENTAGE ERRORS [%]												
		ANALYZED TIME HORIZON [days]												
		1	2	3	4	5	6	7	8	9	10	11	12	13
Number of bus stop	1	0,0	0,0	0,0	1,8	0,0	0,0	0,0	0,0	0,0	0,0	0,0	0,0	0,0
	2	0,0	0,0	0,0	1,8	0,0	0,0	0,0	0,0	0,0	0,0	0,0	0,0	0,0
	3	0,0	0,5	0,5	2,3	1,0	0,0	0,0	0,0	0,0	0,5	0,5	0,0	0,0
	4	0,5	2,5	0,5	3,3	0,5	1,4	0,0	0,5	1,5	1,5	3,5	0,0	0,0
	5	2,0	2,2	1,0	3,8	1,0	4,2	2,1	2,5	2,5	3,0	2,2	0,5	0,7
	6	6,5	9,5	3,5	11,3	7,2	9,4	2,1	8,2	10,4	9,5	12,4	5,5	2,4
	7	8,5	11,7	4,7	12,3	8,2	5,2	20,8	7,5	10,0	9,5	12,3	7,7	5,2
	8	9,2	10,9	5,5	14,0	11,2	5,9	6,3	9,5	16,3	14,5	15,7	8,5	7,3
	9	10,2	11,4	7,2	12,2	12,4	9,7	7,6	11,2	19,5	18,2	15,8	10,7	7,3
	10	11,7	12,2	8,0	17,3	13,2	20,0	6,3	15,0	23,8	25,2	18,7	9,7	9,7
	11	19,0	21,6	12,2	23,7	24,9	23,1	14,2	21,7	24,2	21,7	23,3	12,9	8,0
	12	22,5	24,4	20,4	15,6	20,6	18,5	21,0	19,5	24,3	19,0	22,6	20,8	13,9
	13	22,8	23,3	25,4	20,6	17,8	13,3	22,4	19,9	23,3	20,9	20,1	20,1	24,5
	14	23,0	24,7	36,8	19,5	21,1	17,5	25,5	18,7	21,4	27,2	18,6	18,7	25,8
	15	24,5	24,0	32,0	21,0	22,9	17,3	19,4	16,6	21,1	25,1	22,4	20,2	26,3
	16	25,8	23,0	27,1	19,3	18,0	19,3	17,8	19,3	22,5	30,1	21,6	18,7	26,9
	17	24,4	23,3	26,8	20,1	19,4	17,9	18,1	18,0	22,2	28,9	19,1	19,0	26,9
	18	19,4	18,7	24,5	19,7	20,9	15,3	14,4	18,1	17,5	30,5	21,1	20,7	17,8
	19	7,5	10,1	10,2	8,1	8,8	12,7	4,5	6,5	11,7	27,3	10,3	10,4	12,2

5. Discussion

The results of the research presented in this article indicate that the application of Markov chains to forecast the bus occupancy level in public transport is entirely justified because it represents, to a reasonable degree, the features of the urban public transport system. Compared to the works mentioned in the literature review, which mostly refer to the problem of passenger flow forecasting in the transport network, the authors' research was strictly focused on forecasting the bus comfort level related to the vehicle occupancy, which can be directly used by travelers to optimize their trips and to change their travel patterns to more environmentally friendly ones. The obtained results can be useful not only for fleet management in the public transport system but also for the development of passenger information systems and trip planning.

Nevertheless, this approach requires further research based on a larger data set sample from an automatic counting system over a longer time horizon. It would also be desirable to determine the influence of other factors on the forecasting effectiveness (season of the year, weather, and specificity of the analysed communication line). The calculated forecast errors are the most significant for interchanging stops, due to the high variability of passenger flows in these places. Therefore, it seems justified to carry out studies with the use of heterogeneous Markov chains, where the transition matrix would be variable depending on the time, type of bus stop or communication line. Such an analysis could be very useful for the practical application and further verification of the proposed approach.

6. Conclusions

The discussed issue in the article concerns the problem of forecasting vehicle occupation in public transport. The analysed issue is particularly important due to the growing problems of congestion and the negative impact of road transport in cities. The methodology presented in the paper and the obtained results can significantly contribute to the development of solutions and systems for a better management as well as a cost and energy consumption optimisation in the public transport system. Current and forecasted information related to bus occupancy, when used correctly in the travel information system, such as ATIS, may have a positive impact on the development of urban mobility patterns by encouraging the use of public transport. In this way, it is possible to support the implementation of a sustainable development postulate in the context of transport.

The transportation system, especially public transport, is an artificial, complex, dynamic and uncertain system. These features influence the internal transport process, which is why proper management is challenging to implement. Therefore, it seems appropriate to use discrete Markov processes to determine the expected occupancy level of vehicles at subsequent departures from a given stop on the line. The presented calculation shows that the thesis is correct and creates an incentive for more in-depth investigations.

Author Contributions: conceptualization, P.W. and D.K.; methodology, P.W. and D.K.; formal analysis, P.W.; investigation, A.S.; resources, J.H.A.; data curation, D.K.; writing—original draft preparation D.K.; writing—review and editing, P.W., D.K., A.S. and J.H.A.; supervision, D.K.

Funding: This research received no external funding.

Conflicts of Interest: The authors declare no conflict of interest.

References

1. Adamski, A. *Inteligentne Systemy Transportowe: Sterowanie, Nadzór i Zarządzanie*; Uczelniane Wydaw. Nauk.-Dydakt. AGH im. S. Staszica: Kraków, Poland, 2003.
2. Bifulco, G.N.; Di Pace, R.; Viti, F. Evaluating the effects of information reliability on travellers' route choice. *Eur. Transp. Res. Rev.* **2013**, *6*, 61–70. [CrossRef]
3. Zito, P.; Amato, G.; Amoroso, S.; Berrittella, M. The effect of Advanced Traveller Information Systems on public transport demand and its uncertainty. *Transportmetrica* **2011**, *7*, 31–43. [CrossRef]
4. Balakrishna, R.; Ben-Akiva, M.; Bottom, J.; Gao, S. Information Impacts on Traveler Behavior and Network Performance: State of Knowledge and Future Directions. In *Advances in Dynamic Network Modeling in Complex Transportation Systems*; Springer: New York, NY, USA, 2013; Volume 2, pp. 193–224.
5. Polydoropoulou, A.; Gopinath, D.A.; Ben-Akiva, M. Willingness to Pay for Advanced Traveler Information Systems SmarTraveler Case Study. *Transp. Res. Rec. J. Transp. Res. Board* **1997**, *1588*, 1–9. [CrossRef]
6. Chorus, C.G.; Molin, E.J.E.; Van Wee, B. Use and Effects of Advanced Traveller Information Services (ATIS): A Review of the Literature. *Transp. Rev.* **2006**, *26*, 127–149. [CrossRef]
7. Blasum, U.; Bussieck, M.R.; Hochstattler, W.; Moll, C.H.; Scheel, H.; Winter, T. Scheduling trams in the morning. *Math. Methods Oper. Res.* **1999**, *49*, 137–148.
8. Haase, K.; Deaulniers, G.; Desrosiers, J. Simultanous vehicle and crew scheduling in urban mass transit systems. *Transp. Sci.* **2001**, *35*, 215–343. [CrossRef]
9. Kidwai, F.A.; Marwah, B.R.; Deb, K.; Karim, M.R. A genetic algorithm based bus scheduling model for Transit network. *Proc. East. Asia Soc. Transp. Stud.* **2005**, *5*, 477–489.
10. Jimenez, F.; Roman, A. Urban bus fleet-to-route assignment for pollutant emissions minimization. *Transp. Res. Part E* **2016**, *85*, 120–131. [CrossRef]
11. Li, L.; Lo, H.K.; Cen, X. Optimal bus fleet management strategy for emissions reduction. *Transp. Res. Part D* **2015**, *41*, 330–347. [CrossRef]
12. Beltran, B.; Carrese, S.; Cipriani, E.; Petrelli, M. Transit network design with allocation of green vehicles: A genetic algorithm approach. *Transp. Res. Part C* **2009**, *17*, 475–483. [CrossRef]
13. Lusby, R.M.; Larsen, J.; Bull, S. A Survey on Robustness in Railway Planning. *Eur. J. Oper. Res.* **2017**, *266*, 1–15. [CrossRef]

14. Gancarz, T. Intelligent transport systems. *Res. Tech. Pap. Pol. Assoc. Transp. Eng. Crac.* **1998**, *27*, 75.
15. Cejrowski, M.; Krych, A.; Pawłowski, M. Support for transport management using BIMBA-BIT and VISUM-PT procedures. *Res. Tech. Pap. Pol. Assoc. Transp. Eng. Crac.* **1998**, *27*, 59–68.
16. Moreira, J.M.; De Sousa, J.F. Planning and control indicators for mass transit companies. *Model. Manag. Transp.* **1999**, *2*, 177.
17. Oziomek, J.; Rogowski, A. The optimal allocation of the buses to the suburban lines in Ostrowiec Świętokrzyski. *Autobusy Tech. Eksploat. Syst. Transp.* **2016**, *4*, 14–19.
18. Oziomek, J.; Rogowski, A. Planning the allocation of the buses to the lines in terms of minimizing fuel consumption based on the example of MPK Ostrowiec Świętokrzyski. *TTS Tech. Transp. Szyn.* **2015**, *12*, 1175–1179.
19. Papierkowski, K. Traffic control in public transport in Scandinavia based on the example of the KON-FRAM system. *Res. Tech. Pap. Pol. Assoc. Transp. Eng. Crac.* **1998**, *27*, 203.
20. Li, J.Q.; Head, K.L. Sustainability provisions in the bus-scheduling problem. *Transp. Res. Part D Transp. Environ.* **2009**, *14*, 50–60. [CrossRef]
21. Mo, Y.; Su, Y. Neural networks based on real-time transit passenger volume prediction. In Proceedings of the 2009 2nd Conference on Power Electronics and Intelligent Transportation System(PEITS), Shenzhen, China, 19–20 December 2009; pp. 303–306.
22. Li, Y. Predict the Volume of Passenger Transport of Railway Based on Grey Markov Chain Model. *Adv. Mater. Res.* **2014**, *1030–1032*, 2069–2072. [CrossRef]
23. Zhao, S.Z.; Ni, T.H.; Wang, Y.; Gao, X.T. A new approach to the prediction of passenger flow in a transit system. *Comput. Math. Appl.* **2011**, *61*, 1968–1974. [CrossRef]
24. Liu, L.; Chen, R.C. A novel passenger flow prediction model using deep learning methods. *Transp. Res. Part C Emerg. Technol.* **2017**, *84*, 74–91. [CrossRef]
25. Li, Y.; Wang, X.; Sun, S.; Ma, X.; Lu, G. Forecasting short-term subway passenger flow under special events scenarios using multiscale radial basis function networks. *Transp. Res. Part C Emerg. Technol.* **2017**, *77*, 306–328. [CrossRef]
26. Zhang, J.; Shen, D.; Tu, L.; Zhang, F.; Xu, C.; Wang, Y.; Tian, C.; Li, X.; Huang, B.; Li, Z. A real-time passenger flow estimation and prediction method for urban bus transit systems. *IEEE Trans. Intell. Transp. Syst.* **2017**, *18*, 3168–3178. [CrossRef]
27. Chen, Q.; Li, W.; Zhao, J. The use of LS-SVM for short-term passenger flow prediction. *Transport* **2011**, *26*, 5–10. [CrossRef]
28. Xue, R.; Sun, D.J.; Chen, S. Short-term bus passenger demand prediction based on time series model and interactive multiple model approach. *Discret. Dyn. Nat. Soc.* **2015**, *2015*, 1–11. [CrossRef]
29. Zhou, C.; Dai, P.; Li, R. The passenger demand prediction model on bus networks. In Proceedings of the IEEE 13th International Conference on Data Mining Workshops, Dallas, TX, USA, 7–10 December 2013; pp. 1069–1076.
30. Ma, Z.; Xing, J.; Mesbah, M.; Ferreira, L. Predicting short-term bus passenger demand using a pattern hybrid approach. *Transp. Res. Part C* **2014**, *39*, 148–163. [CrossRef]
31. Tsai, T.H.; Lee, C.K.; Wei, C.H. Neural network based temporal feature models for short-term railway passenger demand forecasting. *Expert Syst. Appl.* **2009**, *36*, 3728–3736. [CrossRef]
32. Wang, Z.; Yang, C.; Zang, C. Short-term passenger flow prediction on bus stop based on the hybrid model. *Adv. Eng. Res.* **2017**, *140*, 343–347.
33. Roos, J.; Bonnevay, S.; Gavin, G. Short-term urban rail passenger flow forecasting: A dynamic bayesian network approach. In Proceedings of the 2016 15th IEEE International Conference on Machine Learning and Applications (ICMLA), Anaheim, CA, USA, 18–20 December 2016; pp. 1034–1039.
34. Zhang, W.; Zhu, J. Passenger traffic forecast based on the Grey-Markov method. In Proceedings of the 2009 IEEE International Conference on Grey Systems and Intelligent Services (GSIS 2009), Nanjing, China, 10–12 November 2009; pp. 630–633.
35. Xiao, Z.S.; Mao, B.H.; Zhang, T. Integrated predicting model for daily passenger volume of rail transit station based on neural network and Markov chain. In Proceedings of the 2018 3rd IEEE International Conference on Cloud Computing and Big Data Analysis (ICCCBDA), Chengdu, China, 20–22 April 2018; pp. 578–583.
36. Altinkaya, M.; Zontul, M. Urban bus arrival time prediction: A review of computational models. *Int. J. Recent Technol. Eng. (IJRTE)* **2013**, *2*, 164–169.

37. Wang, D.; Sun, X.; Li, Y. Using Markov process for passenger structure prediction within comprehensive transportation channel. *J. Comput.* **2013**, *8*, 1072–1077. [CrossRef]
38. Celikoglu, H.B.; Cigizoglu, H.K. Public transportation trip flow modeling with generalized regression neural networks. *Adv. Eng. Softw.* **2007**, *38*, 71–79. [CrossRef]
39. Rudnicki, A. *Jakość Komunikacji Miejskiej*; Zeszyty Naukowo-Techniczne Oddziału SITK w Krakowie: Seria, Brunei, 1999; p. 71.

© 2019 by the authors. Licensee MDPI, Basel, Switzerland. This article is an open access article distributed under the terms and conditions of the Creative Commons Attribution (CC BY) license (http://creativecommons.org/licenses/by/4.0/).

Article

On Cauchy's Interlacing Theorem and the Stability of a Class of Linear Discrete Aggregation Models Under Eventual Linear Output Feedback Controls

Manuel De la Sen

Institute of Research and Development of Processes IIDP, University of the Basque Country, Campus of Leioa, Leioa, PO Box 48940, Bizkaia, Spain; manuel.delasen@ehu.eus

Received: 8 May 2019; Accepted: 21 May 2019; Published: 24 May 2019

Abstract: This paper links the celebrated Cauchy's interlacing theorem of eigenvalues for partitioned updated sequences of Hermitian matrices with stability and convergence problems and results of related sequences of matrices. The results are also applied to sequences of factorizations of semidefinite matrices with their complex conjugates ones to obtain sufficiency-type stability results for the factors in those factorizations. Some extensions are given for parallel characterizations of convergent sequences of matrices. In both cases, the updated information has a Hermitian structure, in particular, a symmetric structure occurs if the involved vector and matrices are complex. These results rely on the relation of stable matrices and convergent matrices (those ones being intuitively stable in a discrete context). An epidemic model involving a clustering structure is discussed in light of the given results. Finally, an application is given for a discrete-time aggregation dynamic system where an aggregated subsystem is incorporated into the whole system at each iteration step. The whole aggregation system and the sequence of aggregated subsystems are assumed to be controlled via linear-output feedback. The characterization of the aggregation dynamic system linked to the updating dynamics through the iteration procedure implies that such a system is, generally, time-varying.

Keywords: Aggregation dynamic system; Discrete system; Epidemic model; Cauchy's interlacing theorem; Output-feedback control; Stability; Antistable/Stable matrix

1. Introduction

Stability and convergence properties are very important topics when dealing with both continuous- and discrete-time controlled dynamic systems. In this context, one of the most important design tools is the closed-loop stabilization of control systems via the appropriate incorporation of stabilizing controllers; see, for instance, [1–4] and references therein. In particular, in [1], and in some references therein, the robust stable adaptive control of tandem of master-slave robotic manipulators using a multi-estimation scheme is discussed. There are several questions of interest in the analysis, such as the fact that the dynamics may be time-varying and imperfectly known, and the fact that a parallel multi-estimation with eventual switching through time is incorporated into the adaptive controller to improve the transient behavior. The speed estimation and stable control of an induction motor based on the use of artificial neural networks is analyzed in [2]. Strategies of decentralized control, including several applications and stabilization tools, are given in [3,4]. In particular, decentralized control is useful when the various subsystems which are integrated in a whole integrated system are located in separate areas, or when the amount of information needed presents difficulties with regards to obtaining completely optimal suitable performance. Thus, the individual controllers associated with the various subsystems get local information about the corresponding subsystems, and eventually some extra partial information about the remaining ones to achieve stabilization, provided that the neglected coupling dynamics are weak enough. Stabilizing decentralized control designs are described in [3] for networked composite systems. Some technical aspects and the results of non-negative matrices of usefulness to describe the properties and behavior of positive dynamic systems, the robustness of matrices against numerical parameterization perturbations of their entries, and the properties of linear dynamic systems are discussed in [5–8].

This paper focuses on the study of sequences of Hermitian matrices of increasing order which are built via block partition aggregation at each iteration, in such a way that both the current iteration and the next one are Hermitian matrices. The basic mathematical tool is the use of the interlacing Cauchy's theorem of the matrix eigenvalues of the matrices of the sequence, which orders the sequences of the eigenvalues as the iteration progresses [8]. Our main objective is to adapt the interlacing theorem in order to use it to derive stability or convergence conditions of the sequence of matrices, and to use the results for the stability of a large-scale discrete aggregation-type dynamic system [9–14]. The paper is organized as follows. Section 2 is devoted to investigating the properties of boundedness and convergence of the sequences of the determinants and the sequences of eigenvalues as the iteration progresses by aggregation of the updated information while maintaining a Hermitian structure. In the particular case when the matrices describing the problem are real, the updated information has a symmetrical structure. The results are used, in particular, to give stability or anti-stability (in the sense that all the matrix eigenvalues of the matrices of the iterative sequence are unstable) conditions to the matrices used in the standard factorization of Hermitian positive definite matrices. Section 3 extends some of the above results to the convergence of sequences of partitioned Hermitian matrices constructed by aggregation of the updated information. Note that the concept of the convergence of matrices is a discrete counterpart of the matrix stability property in the continuous-time domain, since matrices are stability matrices if all their eigenvalues are in the open complex left-half plane. The basic idea that complex square matrices are convergent if their eigenvalues are within the open unit circle centered at zero is taken into account. An example is discussed concerning a SIR epidemic model with contagions between populations of adjacent clusters in Section 4. Section 5 is devoted to developing an application for the stability of an aggregation discrete-time dynamic controlled system whose order increases by successive incorporations of new subsystems as the iteration index progresses, and whose structure keeps a symmetry. Finally, some conclusions are presented at the end the paper. The relevant mathematical proofs are given in the appendix in order to facilitate a direct reading of the manuscript. The system is assumed to be parameterized by real parameters and controlled by linear output-feedback control laws; it is also assumed that the former whole aggregation system and each new aggregated subsystem at each iteration might eventually be coupled.

Notation and Mathematical Symbols

If M is a square Hermitian matrix, then $M > 0$ denotes that it is positive definite and $M \succeq 0$ denotes it is positive semidefinite. Also, $M < 0$ denotes, that it is negative definite.

$$\mathbf{Z}_{0+} = \{z \in \mathbf{Z} : z \geq 0\}; \ \mathbf{Z}_{+} = \{z \in \mathbf{Z} : z > 0\},$$

$$\mathbf{R}_{0+} = \{z \in \mathbf{R} : z \geq 0\}; \ \mathbf{R}_{+} = \{z \in \mathbf{R} : z > 0\},$$

$$\mathbf{C}_{0+} = \{z \in \mathbf{C} : \operatorname{Re} z \geq 0\}; \ \mathbf{C}_{+} = \{z \in \mathbf{C} : \operatorname{Re} z > 0\},$$

$$\overline{n} = \{1, 2, \ldots, n\}$$

I is an identity matrix specified by I_n if it denotes the n-th identity matrix, $A > 0$ denotes that the square matrix A is positive definite (positive semidefinite), $A < 0$ denotes that the square matrix A is negative definite (respectively, negative semidefinite), $A > B$, $A \succeq B$, $A < B$, $A \preceq B$ denote, respectively, that $A - B > 0$, $A - B \succeq 0$, $A - B < 0$ and $A - B \preceq 0$, $\lambda_{\min}(M)$ and $\lambda_{\max}(M)$ denote, respectively, the minimum and maximum eigenvalue of a square real symmetric matrix M, $r(M)$ is the spectral radius of any square complex matrix M, $sp(M)$ is the set of eigenvalues of the Hermitian matrix M. If such a set is ordered with respect to the partial order relation "$\leq$" then the ordered spectrum is denoted by $sp_\leq(M)$. The superscripts * and T stand, respectively, for complex conjugates or transposes of any vector or matrix, $A \otimes B$ is the Kronecker product of the matrices A, if $A \in \mathbf{C}^{n \times m}$ then its vectorization is a vector $vec(A) \in \mathbf{C}^{n \times m}$ whose components are all the rows of A written in column in its order and respecting the order of its respective entries, $A^\dagger$ is the Moore-Penrose pseudoinverse of the matrix A.

2. Technical Results on Partitioned Hermitian Matrices, Cauchy's Interlacing Theorem and Stability

The subsequent result relies on the conditions for the non-singularity of a partitioned Hermitian matrix of order $(n+1)$ which is built by aggregation from a principal Hermitian sub-matrix of order n. Mathematical proof is given in Appendix A.

Lemma 1. *Consider the partitioned matrix* $M\prime = \begin{bmatrix} M & m \\ m^* & d+\widetilde{d} \end{bmatrix} \in \mathbf{C}^{(n+1) \times (n+1)}$ *for any* $n \in \mathbf{Z}_+$, *where* $M \in \mathbf{C}^{n \times n}$ *is Hermitian,* $m \in \mathbf{C}^n$ *and* $d, \widetilde{d} \in \mathbf{R}$. *Then,*

(i) $M\prime$ *is non-singular if and only if* $\det\begin{bmatrix} M & m \\ m^* & d \end{bmatrix} \neq -\widetilde{d} \det M$, *equivalently, if and only if* $\det\begin{bmatrix} M & m \\ m^* & 0 \end{bmatrix} \neq -\det\begin{bmatrix} M & 0 \\ m^* & d+\widetilde{d} \end{bmatrix}$.

(ii) *Assume that* $M \succ 0$ *and* $d > 0$. *Then,* $M\prime \succ 0$ *if and only if* $\det\begin{bmatrix} M & m \\ m^* & d \end{bmatrix} > -\det\begin{bmatrix} M & 0 \\ m^* & \widetilde{d} \end{bmatrix}$. *If* $M \succ 0$ *and* $d \geq 0$ *then* $M\prime \succeq 0$ *if* $\det\begin{bmatrix} M & m \\ m^* & d \end{bmatrix} \geq 0$. *If* $M \prec 0$ *and* $d \leq 0$ *then* $M\prime \preceq 0$ *if* $\det\begin{bmatrix} M & m \\ m^* & d \end{bmatrix} \leq 0$.

The subsequent result relies on some conditions which guarantee the boundedness of the determinant and eigenvalues of a recursive sequence of Hermitian matrices which were obtained and supported by Lemma 1 and Cauchy's interlacing theorem.

Lemma 2. *Consider the recursive sequence of Hermitian matrices* $\{M^{(n)}\}_{n=n_0}^\infty$ *for a given initial* $M^{(n_0)} \in \mathbf{C}^{n_0 \times n_0}$ *for some given arbitrary* $n_0 \in \mathbf{Z}_+$, *where* $M^{(n)} \in \mathbf{C}^{n \times n}$; $\forall n (\geq n_0) \in \mathbf{Z}_+$, *defined by* $M^{(n+1)} = \begin{bmatrix} M^{(n)} & m^{(n)} \\ m^{(n)*} & d^{(n)} + \widetilde{d}^{(n)} \end{bmatrix}$; $\forall n (\geq n_0) \in \mathbf{Z}_+$ *and assume that there is a real sequence* $\{\varepsilon^{(n)}\}_{n=n_0}^\infty \subset [0, 1)$ *such that* $\frac{1}{k_M^{(n)} k_{\widetilde{d}}^{(n)}} \sqrt{m^{(n)*} m^{(n)}} \leq \varepsilon^{(n)}$, *equivalently* $k_M^{(n)} \geq \frac{1}{k_{\widetilde{d}}^{(n)} \varepsilon^{(n)}} \sqrt{m^{(n)*} m^{(n)}}$; $\forall n (\geq n_0) \in \mathbf{Z}_+$, *where* $k_M^{(n)} = |\lambda_{min}(M^{(n)})| \leq K_M^{(n)} = |\lambda_{max}(M^{(n)})|$; $|d^{(n)}| \in [k_d^{(n)}, K_d^{(n)}]$; $|\widetilde{d}^{(n)}| \in [k_{\widetilde{d}}^{(n)}, K_{\widetilde{d}}^{(n)}]$; $\forall n (\geq n_0) \in \mathbf{Z}_+$ *with* $K_d^{(n)} \geq K_{\widetilde{d}}^{(n)}$ *and* $k_d^{(n)} \geq k_{\widetilde{d}}^{(n)}$; $\forall n (\geq n_0) \in \mathbf{Z}_+$. *Then, the following properties hold:*

(i) $\lim\sup_{n \to \infty} (|\det M^{(n+1)}| - |\det M^{(n)}|) \leq 0$, $|\det M^{(n+1)}| \leq |\det M^{(n)}|$ *for any given* $n (\geq n_0) \in \mathbf{Z}_+$ *if* $d^{(n)}$, $\widetilde{d}^{(n)}$ *and* $m^{(n)}$ *satisfy the constraint* $K_d^{(n)} \leq \frac{1 - K_d^{(n)}}{1 + (2^{n+1} - 1)\varepsilon^{(n)}}$; $\forall n (\geq n_0) \in \mathbf{Z}_+$ *with* $K_d^{(n)} \leq 1$, *which becomes* $|\widetilde{d}^{(n)}| \leq \frac{1 - |d^{(n)}|}{1 + (2^{n+1} - 1)\varepsilon^{(n)}}$; $\forall n (\geq n_0) \in \mathbf{Z}_+$ *if* $|d^{(n)}| = K_d^{(n)} = k_d^{(n)} \leq 1$ *and* $|\widetilde{d}^{(n)}| = K_{\widetilde{d}}^{(n)} = k_{\widetilde{d}}^{(n)}$; $\forall n (\geq n_0) \in \mathbf{Z}_+$.

(ii) *Assume that* $sp_\leq(M^{(n)}) = \{\lambda_1^{(n)}, \lambda_2^{(n)}, \ldots, \lambda_n^{(n)}\}$ *and that* $K_d^{(n)} \leq \frac{1 - K_d^{(n)}}{1 + (2^{n+1} - 1)\varepsilon^{(n)}}$ *with* $K_d^{(n)} \leq 1$, $\forall n (\geq n_0) \in \mathbf{Z}_+$, *for some given* $n (\geq n_0) \in \mathbf{Z}_+$. *Then, the following relations hold:*

$$\left|\lambda_{n+1}^{(n+1)}\right| \leq \left|\frac{\prod_{i=1}^n \lambda_i^{(n)}}{\prod_{i=1}^n \lambda_i^{(n+1)}}\right| = \frac{|\det M^{(n)}|}{\left|\prod_{i=1}^n \lambda_i^{(n+1)}\right|} \tag{1}$$

If, furthermore, $M^{(n)} \geq 0$ and $M^{(n+1)} \geq 0$ for some $n(\geq n_0) \in \mathbf{Z}_+$ then

$$\lambda_n^{(n)} \leq \lambda_{n+1}^{(n+1)} \leq \frac{\prod_{i=1}^n \lambda_i^{(n)}}{\prod_{i=1}^n \lambda_i^{(n+1)}}; \ 1 \leq \frac{\lambda_{n+1}^{(n+1)}}{\lambda_n^{(n)}} \leq \frac{\prod_{i=1}^{n-1} \lambda_i^{(n)}}{\prod_{i=1}^n \lambda_i^{(n+1)}}; \ \frac{\prod_{i=2}^{n+1} \lambda_i^{(n+1)}}{\prod_{i=2}^n \lambda_i^{(n)}} \leq \frac{\lambda_1^{(n)}}{\lambda_1^{(n+1)}} \leq 1 \quad (2)$$

(iii) Assume that the constraints of Property (ii) hold with $M^{(n_0)} \geq 0$; $\forall n(\geq n_0) \in \mathbf{Z}_+$ and, furthermore, $\limsup_{n \to \infty} \left(d^{(n)} + \widetilde{d}^{(n)}\right) \leq 1$ and $m^{(n)} = \min\left(o\left(\|M^{(n)}\|\right), o\left(\left|\widetilde{d}^{(n)}\right|\right)\right)$, which is guaranteed if $m^{(n)} = \min\left(o\left(K_M^{(n)}\right), o\left(K_{\widetilde{d}}^{(n)}\right)\right)$. Then, $M^{(n)} \geq 0$ and $M^{(n+1)} \geq 0$; $\forall n(\geq n_0) \in \mathbf{Z}_+$, $\left\{\det M^{(n)}\right\}_{n=n_0}^\infty$ is bounded and the sequence $\left\{sp(M^{(n)})\right\}_{n=n_0}^\infty$ is bounded, if $\det M^{(n_0)}$ is finite, and then $\limsup_{n \to \infty} \left(\det M^{(n+1)} - \det M^{(n)}\right) \leq 0$.

Remark 1. *Concerning Lemma 2 (i), we can focus on the following particular cases of interest A:*

(a) $m^{(n)} = 0$ and $\left|\det M^{(n+1)}\right| \leq \left|\det M^{(n)}\right|$ fails for all $n(\geq \bar{n}_0) \in \mathbf{Z}_+$ and some $\bar{n}_0(\geq n_0) \in \mathbf{Z}_+$. Then, $\left|\det M^{(n+1)}\right| = \left|\det M^{(n)}\right| \left|d^{(n)} + \widetilde{d}^{(n)}\right| > \left|\det M^{(n)}\right|$ so that $K_d^{(n)} + K_{\widetilde{d}}^{(n)} \geq \left|d^{(n)} + \widetilde{d}^{(n)}\right| > 1$ and $\frac{1 - K_d^{(n)}}{1 + (2^{n+1} - 1)\varepsilon^{(n)}} \geq K_{\widetilde{d}}^{(n)} > 1 - K_d^{(n)}$ so that $1 \leq 1 + (2^{n+1} - 1)\varepsilon^{(n)} < 1$, a contradiction. Thus, one has $\left|\det M^{(n+1)}\right| \leq \left|\det M^{(n)}\right|$ for any $n(\geq \bar{n}_0) \in \mathbf{Z}_+$ such that $m^{(n)} = 0$ and also $\limsup_{n \to \infty} \left(\left|\det M^{(n+1)}\right| - \left|\det M^{(n)}\right|\right) \leq 0$.

(b) $d^{(n)} + \widetilde{d}^{(n)} = 0$ and $\left|\det M^{(n+1)}\right| \leq \left|\det M^{(n)}\right|$ fails for all $n(\geq \bar{n}_0) \in \mathbf{Z}_+$ and some $\bar{n}_0(\geq n_0) \in \mathbf{Z}_+$. Note from the definition of the recursive sequence $\left\{M^{(n)}\right\}_{n=n_0}^\infty$ that

$$\det M^{(n+1)} = \det \begin{bmatrix} M^{(n)} & 0 \\ m^{(n)*} & d^{(n)} \end{bmatrix} + \det \begin{bmatrix} M^{(n)} & m^{(n)} \\ m^{(n)*} & \widetilde{d}^{(n)} \end{bmatrix}$$

$$= d^{(n)} \det M^{(n)} + \det \left(\begin{bmatrix} M^{(n)} & 0 \\ 0 & \widetilde{d}^{(n)} \end{bmatrix} \left[I_{n+1} + \begin{bmatrix} M^{(n)^{-1}} & 0 \\ 0 & 1/\widetilde{d}^{(n)} \end{bmatrix} \begin{bmatrix} 0 & m^{(n)} \\ m^{(n)*} & 0 \end{bmatrix} \right] \right) \quad (3)$$

$$= -\widetilde{d}^{(n)} \det M^{(n)} \left(1 - \det \left(I_{n+1} + \begin{bmatrix} M^{(n)^{-1}} & 0 \\ 0 & -1/d^{(n)} \end{bmatrix} \begin{bmatrix} 0 & m^{(n)} \\ m^{(n)*} & 0 \end{bmatrix} \right) \right); \forall n(\geq \bar{n}_0) \in \mathbf{Z}_+$$

Since $\frac{1}{k_M^{(n)} k_{\widetilde{d}}^{(n)}} \sqrt{m^{(n)*} m^{(n)}} \leq \varepsilon^{(n)}$ and $\left\{\varepsilon^{(n)}\right\}_{n=n_0}^\infty \to 0$ then

$$\limsup_{n \to \infty} \left(\left|\det M^{(n+1)}/\det M^{(n)}\right| - \left|\widetilde{d}^{(n)}\right| \left|1 - \det \left(I_{n+1} + \begin{bmatrix} 0 & M^{(n)^{-1}} m^{(n)} \\ -(1/d^{(n)}) m^{(n)*} & 0 \end{bmatrix} \right)\right| \right)$$

$$= \limsup_{n \to \infty} \left|\det M^{(n+1)}/\det M^{(n)}\right| = 0$$

and $\limsup_{n \to \infty} \left(\left|\det M^{(n+1)}\right| - \left|\det M^{(n)}\right|\right) \leq 0$ since, otherwise, if $\limsup_{n \to \infty} \left(\left|\det M^{(n+1)}\right| - \left|\det M^{(n)}\right|\right) > 0$ then $\limsup_{n \to \infty} \left|\det M^{(n+1)}/\det M^{(n)}\right| \in (-\infty, -1) \cup (1, +\infty)$, a contradiction to $\limsup_{n \to \infty} \left|\det M^{(n+1)}/\det M^{(n)}\right| = 0$. Note that if $\widetilde{d}^{(n)} = 0$ then $\left|d^{(n)}\right| \leq 1$ under the given constraints so that if $\left\{m^{(n)}\right\}_{n=n_0}^\infty \to 0$;

(c) $\forall n(\geq \bar{n}_0) \in \mathbf{Z}_+$ and some $\bar{n}_0 \geq n_0$ then $\limsup_{n \to \infty} \left(\left|\det M^{(n+1)}\right| - \left|\det M^{(n)}\right|\right) \leq 0$.

Now, one gets from Lemma 2 [(ii), (iii)] the subsequent dual result concerning the recursion obtained from the inverse of $M^{(n)}$. The use of this result will make it possible to give sufficiency-type conditions regarding the non-singularity of the recursive calculation for any positive integer n, and also as n tends to infinity.

Lemma 3. *For some given arbitrary $n_0 \in \mathbf{Z}_+$ and all $n(\geq n_0) \in \mathbf{Z}_+$, define:*

$$\overline{M}^{(n)} = M^{(n)-1}\left(I_n + m^{(n)}\left(d^{(n)} + \widetilde{d}^{(n)} - m^{(n)*}M^{(n)-1}m^{(n)}\right)^{-1} m^{(n)*}M^{(n)-1}\right) \quad (4)$$

$$\overline{m}^{(n)} = -M^{(n)-1}m^{(n)}\left(d^{(n)} + \widetilde{d}^{(n)} - m^{(n)*}M^{(n)-1}m^{(n)}\right)^{-1} \quad (5)$$

$$\overline{d}^{(n)} + \overline{\widetilde{d}}^{(n)} = \left(d^{(n)} + \widetilde{d}^{(n)} - m^{(n)*}M^{(n)-1}m^{(n)}\right)^{-1} \quad (6)$$

and assume that:

(1) *there is a real sequence $\{\bar{\varepsilon}^{(n)}\}_{n=n_0}^{\infty} \subset [0,1)$ such that $\frac{1}{\bar{k}_M^{(n)}(n)\bar{k}_d^{(n)}}\sqrt{\overline{m}^{(n)*}\overline{m}^{(n)}} \leq \bar{\varepsilon}^{(n)}; \forall n(\geq n_0) \in \mathbf{Z}_+$, where*

$$\bar{k}_M^{(n)} = \left|\lambda_{\min}\left(\overline{M}^{(n)}\right)\right| \leq \overline{K}_M^{(n)} = \left|\lambda_{\max}\left(\overline{M}^{(n)}\right)\right|; \ \overline{d}^{(n)} \in \left[\bar{k}_d^{(n)}, \overline{K}_d^{(n)}\right];$$

$$\overline{\widetilde{d}}^{(n)} \in \left[\bar{k}_{\widetilde{d}}^{(n)}, \overline{K}_{\widetilde{d}}^{(n)}\right]; \ \forall n(\geq n_0) \in \mathbf{Z}_+,$$

$$\widetilde{d}^{(n)} \neq m^{(n)*}M^{(n)-1}m^{(n)} - d^{(n)}$$

(2) $\overline{M}^{(n_0)} \geq 0$, $\limsup_{n \to \infty}\left(\overline{d}^{(n)} + \overline{\widetilde{d}}^{(n)}\right) \leq 1$ and $\overline{m}^{(n)} = \min\left(o\left(\|\overline{M}^{(n)}\|\right), o\left(\overline{\widetilde{d}}^{(n)}\right)\right)$, which is guaranteed if $\overline{m}^{(n)} = \min\left(o\left(\overline{K}_M^{(n)}\right), o\left(\overline{K}_{\widetilde{d}}^{(n)}\right)\right)$. Then, $\overline{M}^{(n)} \geq 0$ and $\overline{M}^{(n+1)} \geq 0; \forall n(\geq n_0) \in \mathbf{Z}_+$, $\left\{\det \overline{M}^{(n)}\right\}_{n=n_0}^{\infty}$ is bounded, the sequence $\left\{sp\left(\overline{M}^{(n)}\right)\right\}_{n=n_0}^{\infty}$ is bounded, if $\det \overline{M}^{(n_0)}$ is finite, and then $\limsup_{n \to \infty}\left(\det\left|\overline{M}^{(n+1)}\right| - \det\left|\overline{M}^{(n)}\right|\right) \leq 0$.

One gets by combining Lemma 2 and Lemma 3 the two subsequent direct results:

Lemma 4. *Assume that $M^{(n_0)} > 0$ for some given arbitrary $n_0 \in \mathbf{Z}_+$ and assume also that the conditions of Lemma 2 (iii) and Lemma 3 hold. Then, $M^{(n+1)} > 0; \forall n(\geq n_0) \in \mathbf{Z}_+$.*

Lemma 5. *Assume that, for some finite $n_0 \in \mathbf{Z}_+$, $A^{(n_0)} \in \mathbf{C}^{n_0 \times n_0}$ is a stability matrix and construct a sequence $\{M^{(n)}\}_{n=n_0}^{\infty}$ according to the recursive rule:*

$$M^{(n+1)} = A^{(n+1)*}A^{(n+1)} = \begin{bmatrix} M^{(n)} & m^{(n)} \\ m^{(n)*} & d^{(n)} + \widetilde{d}^{(n)} \end{bmatrix} = \begin{bmatrix} A^{(n)*}A^{(n)} & m^{(n)} \\ m^{(n)*} & d^{(n)} + \widetilde{d}^{(n)} \end{bmatrix}; \ \forall n \geq n_0$$

with initial condition $M^{(n_0)} = A^{(n_0)*}A^{(n_0)} > 0$. Assume also that $\{M^{(n)}\}_{n=n_0}^{\infty}$ and the sequence of its inverses satisfy the constraints of Lemma 2 [(ii),(iii)] and Lemma 3.

Then, $\{A^{(n)}\}_{n=n_0}^{\infty}$ is a sequence of stability matrices.

The above result can be directly extended for the case when $A^{(n_0)} \in \mathbf{C}^{n_0 \times n_0}$ is antistable, that is, when all its eigenvalues have positive real parts and $M^{(n_0)} > 0$. Then, by using similar arguments, as in the proof of Lemma 5 based on the continuity of the matrix eigenvalues with respect to its entries and supported by Lemmas 2,3, according to Cauchy's interlacing theorem, one concludes that $\{A^{(n)}\}_{n=n_0}^{\infty}$ consists of antistable members.

Lemma 6. *Lemma 5 holds "mutatis-mutandis" if $A^{(n_0)} \in \mathbf{C}^{n_0 \times n_0}$ is antistable.*

3. Some Extended Results Related to Sequences of Convergent Matrices

In order to be able to adapt the above results to discrete dynamic systems, the well-known result that that the stability domain of a convergent matrix (i.e., a "stable" discrete matrix) is the open unit circle of the complex plane centered at zero has to be taken into account. Note that, in particular, $A \in \mathbf{C}^{n \times n}$ is convergent if $sp(A) \subset C_1 = \{z \in \mathbf{C} : |z| < 1\}$ so that $\{A^m\}_0^\infty \to 0$ as $m \to \infty$. It turns out that convergent matrices describe the stability property in the discrete sense. In other words, the solution of the discrete difference vector equation $z_{m+1} = Az_m$, where $A \in \mathbf{C}^{n \times n}$, converges to $0 \in \mathbf{R}^n$ for any given $z_0 \in \mathbf{C}^n$ if and only if $A \in \mathbf{C}^{n \times n}$ is convergent. The relevant results of Section 2 can be extended to this situation as follows, provided that $A \in \mathbf{C}^{n \times n}$ is also Hermitian. Consider the following cases:

Case a: $sp(A) \subset C_{1+} = \{z \in \mathbf{C} : 0 < z < 1\} \subset C_1$ so that $A (\in \mathbf{C}^{n \times n}) > 0$. Then, it is convergent if and only if $(I_n - A) > 0$. The proof is direct since if $(I_n - A) > 0$ then $x^*Ax < x^*x$ for any $x(\neq 0) \in \mathbf{C}^n$. Thus, by taking any $\lambda(\neq 0) \in sp(A)$ of eigenvector $x \in \mathbf{C}^n$, one determines that $\lambda < 1$ if $\lambda \neq 0$ and $\lambda = 0$ directly fulfills the constraint. This proves the sufficiency part. The "only if part" follows, since if $(I_n - A) > 0$ fails, there is $\lambda(\neq 0) \in sp(A)$ of eigenvector $x \in \mathbf{C}^n$ such that $x^*Ax \geq x^*x$ then $\lambda \geq 1$ and A is not convergent.

Case b: $sp(A) \subset C_{1-} = \{z \in \mathbf{C} : -1 < z < 0\} \subset C_1$ so that $A (\in \mathbf{C}^{n \times n}) < 0$. Then, it is convergent if and only if $(I_n + A) > 0$. The proof is direct, since if $(I_n + A) > 0$, then $x^*Ax > -x^*x$ for any $x(\neq 0) \in \mathbf{C}^n$. Thus, by taking any $\lambda(\neq 0) \in sp(A)$ of eigenvector $x \in \mathbf{C}^n$, one determines that $0 > \lambda > -1$. The remainder of the proof follows Case a closely.

Case c: $sp(A) \subset C_1$ so that $A^2 (\in \mathbf{C}^{n \times n}) > 0$ so that $sp(A^2) \subset C_{1+}$. Then, it is convergent if and only if $(I_n - A^2) > 0$ according to *Case a* by replacing $A \to A^2$. Note that *Case c* is included *Case a* and *Case b*.

Now, for *Case a*, replace $M^{(n+1)}$, defined in Lemma 2, by $I_{n+1} - M^{(n+1)} = \begin{bmatrix} I_n - M^{(n)} & -m^{(n)} \\ -m^{(n)*} & 1 - (d^{(n)} + \tilde{d}^{(n)}) \end{bmatrix}$ and it has to be guaranteed that if $M^{(n)}$ is Hermitian, then $(I_n - M^{(n)})$ is also Hermitan, and $(I_n - M^{(n_0)}) > 0$ for some $n_0 \in \mathbf{Z}_+$ then $(I_{n+1} - M^{(n+1)}) > 0; \forall n \geq n_0$.

For *Case b*, replace $M^{(n+1)} \to (I_{n+1} + M^{(n+1)}) = \begin{bmatrix} I_n + M^{(n)} & m^{(n)} \\ m^{(n)*} & 1 + d^{(n)} + \tilde{d}^{(n)} \end{bmatrix}$ and it has to be guaranteed that if $M^{(n)}$ is Hermitian, then $(I_n + M^{(n)})$ is also Hermitan, and $(I_n + M^{(n_0)}) > 0$ for some $n_0 \in \mathbf{Z}_+$ then $(I_{n+1} + M^{(n+1)}) > 0; \forall n \geq n_0$.

For *Case c*, note that $(I_n - M^{(n)^2}) = (I_n + M^{(n)})(I_n - M^{(n)})$ so that

$$(I_n - M^{(n)^2}) = \begin{bmatrix} I_n + M^{(n)} & m^{(n)} \\ m^{(n)*} & 1 + d^{(n)} + \tilde{d}^{(n)} \end{bmatrix} \begin{bmatrix} I_n - M^{(n)} & -m^{(n)} \\ -m^{(n)*} & 1 - (d^{(n)} + \tilde{d}^{(n)}) \end{bmatrix}$$

then, replace

$$M^{(n+1)} \to (I_{n+1} - M^{(n+1)^2}) = \begin{bmatrix} I_n - M^{(n)2} & -((d^{(n)} + \tilde{d}^{(n)})I_n + M^{(n)})m^{(n)} \\ -m^{(n)*}((d^{(n)} + \tilde{d}^{(n)})I_n + M^{(n)*}) & 1 - (d^{(n)} + \tilde{d}^{(n)})^2 - m^{(n)*}m^{(n)} \end{bmatrix}$$

and it has to be guaranteed that if $M^{(n)^2}$ is Hermitian and $(I_n - M^{(n_0)^2}) > 0$ for some $n_0 \in \mathbf{Z}_+$ then $(I_{n+1} - M^{(n+1)^2}) > 0; \forall n \geq n_0$. Since $A > 0$ is Hermitian, it is of the form $A = E^*E$ for some full rank n-matrix E. Then, $A^2 = (E^*E)^2 = E^*EE^*E = A^*A$. If $A < 0$ then $(-A) > 0$ and $(-A)^2 = (F^*F)^2 = F^*FF^*F = (-A^*)(-A) = A^*A$ for some full rank n-matrix F. Then, Cases a and b can be dealt with using *Case c* by replacing $A^2 \to A^*A$.

By taking advantage from the fact that a complex square matrix A is convergent (i.e., stable in the discrete sense) if and only if the Hermitian matrix A^*A is convergent, we now build a sequence $\{M^{(n)}\}_{n=0}^{\infty}$ of Hermitian matrices as follows, in order to discuss the convergence of its members, provided that $M^{(n_0)}$ is convergent for some given $n_0 \in \mathbb{Z}_{0+}$ or, with no loss in generality, provided that $M^{(0)}$ is convergent. Then,

$$M^{(n+1)} = \begin{bmatrix} M^{(n)} & \lambda^{(n)}m^{(n)} \\ \lambda^{(n)}m^{(n)*} & \delta^{(n)} \end{bmatrix} = \begin{bmatrix} M^{(n)} & 0 \\ 0 & \delta^{(n)} \end{bmatrix} + \begin{bmatrix} 0 & \lambda^{(n)}m^{(n)} \\ \lambda^{(n)}m^{(n)*} & 0 \end{bmatrix}; \forall n \in \mathbb{Z}_{0+} \quad (7)$$

with $\{\lambda^{(n)}\}_{n=0}^{\infty} \subset [0,1)$. Then,

$$\left|\lambda_{\max}\begin{bmatrix} M^{(n)} & \lambda^{(n+1)}m^{(n)} \\ \lambda^{(n)}m^{(n)*} & \delta^{(n)} \end{bmatrix}\right| \leq \left|\lambda_{\max}\begin{bmatrix} M^{(n)} & 0 \\ 0 & \delta^{(n)} \end{bmatrix}\right| + \left|\lambda_{\max}\begin{bmatrix} 0 & \lambda^{(n)}m^{(n)} \\ \lambda^{(n)}m^{(n)*} & 0 \end{bmatrix}\right|$$
$$= \max(|\lambda_{\max}(M^{(n)})|, |\delta^{(n)}|) + \lambda^{(n)}\sqrt{m^{(n)*}m^{(n)}} \quad (8)$$
$$\leq \max(1-\varepsilon^{(n)}, |\delta^{(n)}|) + \lambda^{(n)}\sqrt{m^{(n)*}m^{(n)}} < 1-\varepsilon^{(n)}; \forall n \in \mathbb{Z}_{0+}$$

which holds if

$$m^{(n)*}m^{(n)} < \frac{1}{\lambda^{(n)2}}(1-\varepsilon^{(n)} - \max(1-\varepsilon^{(n)}, |\delta^{(n)}|))^2; \forall n \in \mathbb{Z}_{0+} \quad (9)$$

or, $\|m^{(n)}\|_2 < \frac{1}{\lambda^{(n)2}}(1-\varepsilon^{(n)} - \max(1-\varepsilon^{(n)}, |\delta^{(n)}|)); \forall n \in \mathbb{Z}_{0+}$, provided that $\varepsilon^{(n+1)} < \varepsilon^{(n)}; \forall n \in \mathbb{Z}_{0+}$, that is $\{\varepsilon^{(n)}\}_{n=0}^{\infty} \subset [0,1)$ is strictly decreasing, so $\{\varepsilon^{(n)}\}_{n=0}^{\infty} \to 0$, and $|\delta^{(n)}| < 1-\varepsilon^{(n+1)}; \forall n \in \mathbb{Z}_{0+}$.

Now, assume that the iterations to build $\{M^{(n+1)}\}_{n=0}^{\infty}$ do not add a new row and column to obtain $M^{(n+1)}$ from $M^{(n)}$ via the contribution of the members of an updating sequence $\{\overline{M}^{(n)}\}_{n=0}^{\infty}; \forall n \in \mathbb{Z}_{0+}$ but a set of the, in general. Then, one may get that:

$$M^{(n+1)} = \begin{bmatrix} M^{(n)} & \lambda^{(n)}\overline{M}^{(n)} \\ \lambda^{(n)}\overline{M}^{(n)*} & \Delta^{(n)} \end{bmatrix} = \begin{bmatrix} M^{(n)} & 0 \\ 0 & \Delta^{(n)} \end{bmatrix} + \begin{bmatrix} 0 & \lambda^{(n)}\overline{M}^{(n)} \\ \lambda^{(n)}\overline{M}^{(n)*} & 0 \end{bmatrix}; \forall n \in \mathbb{Z}_{0+} \quad (10)$$

so that

$$\left|\lambda_{\max}M^{(n+1)}\right| = \left|\lambda_{\max}\begin{bmatrix} M^{(n)} & \lambda^{(n)}\overline{M}^{(n)} \\ \lambda^{(n)}\overline{M}^{(n)*} & \Delta^{(n)} \end{bmatrix}\right| \leq \left|\lambda_{\max}\begin{bmatrix} M^{(n)} & 0 \\ 0 & \Delta^{(n)} \end{bmatrix}\right| + \left|\lambda_{\max}\begin{bmatrix} 0 & \lambda^{(n)}\overline{M}^{(n)} \\ \lambda^{(n)}\overline{M}^{(n)*} & 0 \end{bmatrix}\right|$$
$$= \max(|\lambda_{\max}(M^{(n)})|, |\lambda_{\max}(\Delta^{(n)})|) + \lambda^{(n)}\lambda_{\max}^{1/2}(\overline{M}^{(n)*}\overline{M}^{(n)}) \quad (11)$$
$$\leq \max(1-\varepsilon^{(n)}, |\lambda_{\max}(\Delta^{(n)})|) + \lambda^{(n)}\lambda_{\max}^{1/2}(\overline{M}^{(n)*}\overline{M}^{(n)}) < 1-\varepsilon^{(n+1)}; \forall n \in \mathbb{Z}_{0+}$$

which holds by complete induction if $M^{(0)}$ is convergent and

$$\|\overline{M}^{(n)}\|_2^2 = \lambda_{\max}(\overline{M}^{(n)*}\overline{M}^{(n)}) < \frac{1}{\lambda^{(n)2}}(1-\varepsilon^{(n+1)} - \max(1-\varepsilon^{(n)}, |\lambda_{\max}(\Delta^{(n)})|))^2; \quad (12)$$
$$\forall n \in \mathbb{Z}_{0+}$$

or, $\|\overline{M}^{(n)}\|_2 < \frac{1}{\lambda^{(n)}}(1-\varepsilon^{(n+1)} - \max(1-\varepsilon^{(n)}, |\lambda_{\max}(\Delta^{(n)})|)); \forall n \in \mathbb{Z}_{0+}$, provided that $\varepsilon^{(n+1)} < \varepsilon^{(n)}$, that is $\{\varepsilon^{(n)}\}_{n=0}^{\infty} \subset [0,1)$ is strictly decreasing, so $\{\varepsilon^{(n)}\}_{n=0}^{\infty} \to 0$, and $|\lambda_{\max}(\Delta^{(n)})| < 1-\varepsilon^{(n+1)}; \forall n \in \mathbb{Z}_{0+}$. This implies that $\{M^{(n)}\}_{n=0}^{\infty}$ is convergent.

In the particular case that for some $\lambda \in [0,1)$, $\lambda^{(n)} = \lambda^n; \forall n \in \mathbb{Z}_{0+}$, such a λ is a forgetting factor of the iteration.

We now consider the matrix factorization $M^{(n)} = A^{(n)*}A^{(n)}$; $\forall n \in \mathbf{Z}_{0+}$. By construction, $M^{(n)}$ is Hemitian (then square), even if $A^{(n)}$ is not square; $\forall n \in \mathbf{Z}_{0+}$. In the case when $A^{(n)}$ is not square, and since its order strictly increases as n increases, it is possible to consider the convergence of the sequence $\{A^{(n)}\}_{n=0}^{\infty}$ (without invoking the values of its eigenvalues) as the following property $\{A^{(n)}\}_{n=0}^{\infty}$ is asymptotically convergent if $\lim_{m\to\infty}\{\|A^{(n+\xi)}\| - \|A^{(n)}\|\|^m\}_{n=0}^{\infty} = 0$ for any given $\xi \in \mathbf{Z}_+$. $\{A^{(n)}\}_{n=0}^{\infty}$ is convergent if $\lim_{m\to\infty}\{A^{(n)^m}\}_{n=0}^{\infty} = 0$. The following related results are direct of simple proofs given in Appendix A:

Lemma 7. *If $\{A^{(n)}\}_{n=0}^{\infty}$ is convergent then it is asymptotically convergent. The inverse is, in general, not true.*

Lemma 8. *Assume that $M^{(n)} = A^{(n)*}A^{(n)}$; $\forall n \in \mathbf{Z}_{0+}$ is a complex square matrix of any arbitrary order. Then:*

(i) *If $\|A^{(n)}\|_2 < 1$; $\forall n \in \mathbf{Z}_{0+}$ then $\{A^{(n)}\}_{n=0}^{\infty}$ and $\{M^{(n)}\}_{n=0}^{\infty}$ are convergent sequences.*

(ii) *If $\|M^{(n)}\|_2 < 1$; $\forall n \in \mathbf{Z}_{0+}$ then $\{M^{(n)}\}_{n=0}^{\infty}$ and $\{A^{(n)}\}_{n=0}^{\infty}$ are convergent sequences.*

(iii) *For any $n \in \mathbf{Z}_{0+}$, $\|A^{(n)}\|_2 < 1$ if and only if $\|M^{(n)}\|_2 < 1$. $\{A^{(n)}\}_{n=0}^{\infty}$ is convergent if and only if $\{M^{(n)}\}_{n=0}^{\infty}$ is convergent.*

4. Example of SIR-Type Epidemic Models of Inter-Community Clusters

The stability of the equilibrium points of the epidemic models is an interesting topic which is of great relevance to healthcare management. See, for instance, [15–19]. Now, we discuss an epidemic based-model related to stabilization under the given framework of the Cauchy's interlacing theorem.

Example 1. *Consider the subsequent continuous-time linearized epidemic model with Q community clusters:*

$$\dot{S}_{i+1} = v_i(1 - \mu_i)I_i - \beta_{i+1}S_{i+1}$$

$$\dot{I}_{i+1} = \beta_{i+1}S_{i+1} - v_{i+1}I_{i+1} \tag{13}$$

$$\dot{R}_{i+1} = v_{i+1}\mu_{i+1}I_i$$

for $i = 0, 1, \ldots, Q-1$ with $S_i(0) = S_{i0} \geq 0$, $I_i(0) = I_{i0} \geq 0$, $R_i(0) = R_{i0} \geq 0$; $\forall i \in \overline{Q}$ are the initial conditions of the susceptible, infectious and recovered subpopulations, respectively, and $I_0(0) = 0$, $v_0 = 1$. In this model, the infectious subpopulation I_i of a community $i \in \overline{Q} = \{1, 2, \ldots, Q\}$ may infect the population of the neighboring community $(i + 1)$. The parameterization is as follows: $\beta_{(.)}$ are the disease transmission rates, $v_{(.)}$ are the removal rates and $\mu_{(.)}$ are the separation constants which bifurcate the disease rate between the local community and the total community. Note that the assumption $\mu_0 = 1$ implies that the first cluster is not affected by contagions from any other cluster, [15]. A simple analysis of the trajectory solution of the first cluster shows that

$$S_1(t) = e^{-\beta_1 t}S_{10} \to 0 \text{ as } t \to \infty$$

at an exponential rate, irrespective of the initial conditions, and is definitively bounded,

$$I_1(t) = e^{-v_1 t}I_{10} + \int_0^t e^{-v_1(t-\tau)}S_1(\tau)d\tau = e^{-v_1 t}\left(I_{10} + S_{10}\int_0^t e^{(v_1-\beta_1)\tau}d\tau\right)$$
$$= \left(e^{-v_1 t}I_{10} + \frac{e^{-\beta_1 t} - e^{-v_1 t}}{v_1 - \beta_1}S_{10}\right) \to 0 \text{ as } t \to \infty$$

if $v_1 \neq \beta_1$. If $v_1 = \beta_1$ then $I_1(t) = e^{-v_1 t}(I_{10} + S_{10}t) \to 0$ as $t \to \infty$. In both cases, the convergence is of exponential order, irrespective of the initial conditions, and is definitively bounded, and

$$R_1(t) = R_{10} + v_1\mu_1 \int_0^t I_i(\tau)d\tau = R_{10} + v_1\mu_1 \int_0^t \left(e^{-v_1\tau}I_{10} + \frac{e^{-\beta_1\tau}-e^{-v_1\tau}}{v_1-\beta_1}S_{10}\right)d\tau$$

$$= R_{10} + v_1\mu_1 \int_0^t \left(e^{-v_1\tau}I_{10} + \frac{e^{-\beta_1\tau}-e^{-v_1\tau}}{v_1-\beta_1}S_{10}\right)d\tau$$

$$= R_{10} + v_1\mu_1 I_{10}\frac{1-e^{-v_1 t}}{v_1} + \frac{v_1\mu_1 S_{10}}{v_1-\beta_1}\left(e^{-\beta_1 t} - e^{-v_1 t}\right) \to R_{10} + \mu_1 I_{10} + \frac{v_1\mu_1 S_{10}}{v_1-\beta_1}\left(e^{-\beta_1 t} - e^{-v_1 t}\right)$$

as $t \to \infty$

if $v_1 \neq \beta_1$ with solution which is definitively bounded, and, if $v_1 = \beta_1$ then

$$R_1(t) = R_{10} + v_1\mu_1 \int_0^t e^{-v_1\tau}(I_{10} + S_{10}\tau)d\tau$$
$$= R_{10} + \mu_1\left[\left(1-e^{-v_1 t}\right)I_{10} + \left(\frac{1}{v_1}\left(1-e^{-v_1 t}\right) - te^{-v_1 t}\right)S_{10}\right] \to R_{10} + \mu_1\left[I_{10} + \frac{S_{10}}{v_1}\right] \text{ as }$$
$$t \to \infty$$

with a solution which is definitively bounded. As a result, the total subpopulation at the first cluster is also definitively bounded, and it converges asymptotically to the limit value of the recovered subpopulation.

The solution trajectory is also definitively nonnegative since the matrix of dynamics $\begin{bmatrix} -\beta_1 & 0 & 0 \\ \beta_1 & -v_1 & 0 \\ 0 & v_1\mu_1 & 0 \end{bmatrix}$ is a Metzler matrix and the initial conditions are non-negative. Interpretation shows that the total equilibrium subpopulation is that of the disease-free equilibrium which only has a recovered subpopulation. It can be surprising at a first glance to see that the usual nonlinear term $\beta_1 S_1(t)I_1(t)$ is the susceptible and infectious subpopulations evolutions of the corresponding SIR Kermack-Mcendrick model counterpart is replaced by a linear term. However, for stability purposes, there is no substantial distinct qualitative behavior between both models, since in this case, $S_1(t) = e^{-\beta_1 \int_0^t I_1(\tau)d\tau}S_{10}$ is strictly decreasing for $t \geq 0$ and $I_1(t) = e^{\int_0^t (\beta_1 S_1(\tau)-v_1)d\tau}I_{10}$ is also strictly decreasing for $t \geq 0$ provided that $S_{10} < \frac{v_1}{\beta_1}$. Now describe the whole model (14) of Nclusters in a more compact way through the individual states $x_i = (S_i, I_i, R_i)^T; i \in \overline{N}$ and associated matrices of self-dynamics for each $i \in \overline{N}$ and coupled dynamics with the respective preceding cluster $(i-1) \in \overline{N}$:

$$A_i = \begin{bmatrix} -\beta_i & 0 & 0 \\ \beta_i & -v_i & 0 \\ 0 & v_i\mu_i & 0 \end{bmatrix}, \forall i \in \overline{N}; A_{i,i-1} = \begin{bmatrix} 0 & v_{i-1}(1-\mu_{i-1}) & 0 \\ 0 & 0 & 0 \\ 0 & 0 & 0 \end{bmatrix}, \quad (14)$$
$$\forall i(\geq 2) \in \overline{N}; A_{0,-1} = 0$$

so that (13) is equivalently described as

$$\dot{x}_i(t) = A_i x_i(t) + A_{i,i-1}x_{i-1}(t), \; x_i(0) = x_{i0}(\geq 0); \; \forall i \in \overline{N} \quad (15)$$

with $x_0(t) \equiv 0$ for $t \geq 0$, and compactly, as follows:

$$\dot{x}(t) = Ax(t), \; x(0) = x_0 \quad (16)$$

where $x(t) = \left(x_1^T(t), x_2^T(t), \cdots, x_N^T(t)\right)^T$ and

$$A = \begin{bmatrix} A_1 & 0 & 0 & \cdots & 0 \\ A_{21} & A_2 & 0 & \cdots & 0 \\ 0 & A_{32} & A_{30} \cdots & \vdots & \vdots & \cdots & 0 \\ \vdots & & \vdots & & & & \\ 0 & 0 & \cdots & \cdots & 0 & A_{N,N-1} & A_N \end{bmatrix} \qquad (17)$$

Thus, system (15), like (16), (17), can be interpreted as an aggregation model given by starting with the first cluster and successively incorporating the dynamics of the remaining clusters. Now, define the symmetric matrix $M = M^{(N)} = A^T A$. Then, define:

$$M^{(1)} = A_1^T A_1$$

$$M^{(2)} = \begin{bmatrix} A_1^T & A_{21}^T \\ 0 & A_2^T \end{bmatrix} \begin{bmatrix} A_1 & 0 \\ A_{21} & A_2 \end{bmatrix} = \begin{bmatrix} A_1^T A_1 + A_{21}^T A_{21} & A_{21}^T A_2 \\ A_2^T A_{21} & A_2^T A_2 \end{bmatrix}$$
$$= \begin{bmatrix} M^{(1)} & A_{21}^T A_2 \\ A_2^T A_{21} & A_2^T A_2 \end{bmatrix} + \widetilde{M}^{(2)}$$

$$M^{(3)} = \begin{bmatrix} A_1^T & A_{21}^T & 0 \\ 0 & A_2^T & A_{32}^T \\ 0 & 0 & A_3^T \end{bmatrix} \begin{bmatrix} A_1 & 0 & 0 \\ A_{21} & A_2 & 0 \\ 0 & A_{32} & A_3 \end{bmatrix} = \begin{bmatrix} M^{(2)} & 0 \\ & A_{32}^T A_3 \\ 0 & A_3^T A_{32} & A_3^T A_3 \end{bmatrix} + \widetilde{M}^{(3)} \qquad (18)$$

$$M^{(N)} = \begin{bmatrix} M^{(N-1)} & m^{(N-1)} \\ m^{(N-1)T} & A_3^T A_3 \end{bmatrix} + \widetilde{M}^{(N)}$$

$$\widetilde{M}^{(2)} = \begin{bmatrix} A_{21}^T A_{21} & 0 \\ 0 & 0 \end{bmatrix}, \widetilde{M}^{(3)} = \begin{bmatrix} A_{21}^T A_{21} & 0 & 0 \\ 0 & A_{32}^T A_{32} & 0 \\ 0 & 0 & 0 \end{bmatrix}$$

$$\widetilde{M}^{(N)} = \begin{bmatrix} A_{21}^T A_{21} & 0 & \cdots & 0 \\ 0 & A_{32}^T A_{32} & 0 \cdots & 0 \\ 0 & \cdots & \ddots & \\ 0 & \cdots & \cdots & A_{N,N-1}^T A_{N,N-1} \\ & & & 0 \end{bmatrix}$$

if $N \geq 2$ and $\widetilde{M}^{(1)} = 0$. By inspection of (18), one concludes that $\|\widetilde{M}^{(n)}\| = 0\left(\sum_{n=2}^{n} \sup^2 \|m^{(n)}\|\right)$ for $i = 2, 3, \ldots, N$ which concludes that, if $\{A_{n-1,n}\}_{n=2}^{N} \to 0$ as $N \to \infty$ in such a way that $\{\|m^{(j)}\|\}_{j=2}^{N} \to 0$ as $N \to \infty$, for instance, if the convergence is at exponential rate, then $\lim_{N \to \infty} \sum_{n=2}^{N} \sup^2 \|m^{(n)}\| \leq K_m < +\infty$. Furthermore, if A_1 is a stability matrix of absolute stability abscissa which are sufficiently larger than K_m, then the dynamic system (16),(17) is globally asymptotically stable according to Lemma 5. In particular, note that if there is any pair of stable complex conjugate eigenvalues $s_{1,2} = \mu \pm iv$ ($\mu < 0$, $v > 0$) for the first cluster, then there is a submatrix of $M^{(1)}$,

$$M^{s(1)} = A_1^{sT} A_1^s = \begin{bmatrix} \mu & -v \\ v & \mu \end{bmatrix} \begin{bmatrix} \mu & v \\ -v & \mu \end{bmatrix} = \begin{bmatrix} \mu^2 + v^2 & 0 \\ 0 & \mu^2 + v^2 \end{bmatrix} = \begin{bmatrix} \lambda_1^{(1)} & 0 \\ 0 & \lambda_2^{(1)} \end{bmatrix}$$

in the real canonical form. Since $\mu^{(1)} = \mu < 0$ then $\lambda_2^{(1)} = \lambda_1^{(1)} = \mu^2 + v^2 > 0$. The maximum and minimum corresponding eigenvalues of $M^{s(2)}$ are no less than $\lambda_2^{(1)}$ and no larger than $\lambda_1^{(1)}$, respectively, from Cauchy's interlacing theorem. Since the eigenvalues are continuous functions of the matrix entries, and since $M^{s(n)}$ is positive definite any critical eigenvalue of a member $M^{s(j)}$ of the sequence, $\{M^{s(n)}\}_{n=2}^{\infty}$ implies a lot of stability of the corresponding A_j^s. This is avoided if $\lim_{N\to\infty} \sum_{n=2}^{N} \sup^2 \|m^{(n)}\| \leq K_m < +\infty$, implying also that $\{m^{(n)}\}_2^N \to 0$ as $N \to \infty$, and the sequence of separation constants $\{\mu_n\}_{n=2}^{N} \to 1$ as $N \to \infty$ if K_m is small enough related to μ. The physical interpretation relies on the fact that the contagion link from a cluster to the next one is weakened sufficiently quickly as the cluster index increases, due to the fact of the numbers of the infected subpopulations are rapidly decreasing as the cluster index increases at a sufficiently large rate.

5. Dynamic Linear Discrete Aggregation Model with Output Delay and Linear Feedback Control

In this section, the convergence results of Section 3 are applied to a dynamic discrete system which is built by the aggregation of discrete dynamic subsystems subject to linear output feedback control. Since we are dealing with a physical system, it turns out that the formalism of Section 2 can be developed by invoking conditions related to real symmetric systems, rather than to complex Hermitian ones, when necessary. It would suffice to describe the state by expressing the matrix of dynamics in the real canonical form and to transform the control and output matrices by the appropriate similarity matrix. The necessary mathematical proof is given in Appendix A.

Consider the aggregation linear discrete dynamic system subject to r point delays under linear output-feedback:

$$x^{0(n+1)} = A^{0(n)}x^{(n)} + \hat{A}^{0(n)}\hat{x}^{(n)} + \sum_{j=1}^{r} B_j^{(n)} y^{(n-j)} + B_0^{(n)} u^{(n)} \tag{19}$$

$$y^{(n)} = C^{(n)} x^{(n)} \tag{20}$$

$$u^{(n)} = \sum_{j=0}^{r_n} K_j^{(n)} y^{(n-j)} = \sum_{j=0}^{r_n} K_j^{(n)} C x^{(n-j)} \tag{21}$$

$\forall n \in \mathbf{Z}_{0+}$, with initial conditions $x^{0(0)} = x_0$, where $\{n_i\}_{i=0}^{\infty}$ is a sequence of positive integer numbers, $x^{(n)} \in \mathbf{R}^{\sum_{i=0}^{n} n_i}$ is the "a priori" vector state at the n-th iteration, $\hat{x}^{(n)} \in \mathbf{R}^{n_n}$ is the aggregated "a priori" new substate at the n-th iteration (that is basically, the new information needed to update the state vector and its dimension) and $x^{0(n+1)} \in \mathbf{R}^{\sum_{i=0}^{n} n_i}$ is the "a priori" whole state at the $(n+1)$-th iteration. Also, $x^{0(n)} \in \mathbf{R}^{\sum_{i=0}^{n-1} n_i}$, $u^{(n)} \in \mathbf{R}^{\sum_{i=0}^{n} m_i}$ and $y^{(n)} \in \mathbf{R}^{\sum_{i=0}^{n} p_i}$ are, respectively, the "a priori" input and measurable output vectors at the n-th iteration and $r \subset \mathbf{Z}_+$ is a sequence of delays influencing the global dynamics. The sequences of matrices of dynamics $\{A^{0(n)}\}_{n=0}^{\infty}$ and $\{\hat{A}^{0(n)}\}_{n=0}^{\infty}$, control $\{B_0^{(n)}\}_{n=0}^{\infty}$, output-state coupling $\{B_j^{(n)}\}_{n=0}^{\infty}$ for $j \in \bar{r}$ are of members $A^{0(n)} \in \mathbf{R}^{(\sum_{i=0}^{n} n_i) \times (\sum_{i=0}^{n} n_i)}$, $\hat{A}^{(n)} \in \mathbf{R}^{(\sum_{i=0}^{n} n_i) \times n_n}$, and $B_0^{(n)} \in \mathbf{R}^{(\sum_{i=0}^{n} n_i) \times (\sum_{i=0}^{n} m_i)}$ and $B_j^{(n)} \in \mathbf{R}^{(\sum_{i=0}^{n} n_i) \times (\sum_{i=0}^{n-j} p_i)}$ for $j \in \bar{r}$ and the output matrix $C^{(n)} \in \mathbf{R}^{(\sum_{i=0}^{n} p_i) \times (\sum_{i=0}^{n} n_i)}$. The sequences of matrices $\{K_j^{(n)}\}_{n=0}^{\infty}$, with $K_j^{(n)} \in \mathbf{R}^{(\sum_{i=0}^{n} m_i) \times (\sum_{i=0}^{n} p_i)}$ for $j \in \bar{r} \cup \{0\}$, are the output-feedback control gains which generate the control law sequence $\{u^{(n)}\}_{n=0}^{\infty}$.

The dynamics of the new dynamics at the $(n+1)$-th iteration aggregated to the former global aggregation system of state $x^{(n)}$ obtained at the n-th iteration, are assumed to be described by:

$$\hat{x}^{(n+1)} = \hat{A}^{(n+1)} x^{(n)} + \left(\hat{D}^{(n)} + \widetilde{D}^{(n)}\right) \hat{x}^{(n)} + \sum_{j=1}^{r} \left(\hat{B}_j^{a(n)} y^{(n-j)} + \hat{B}_j^{(n)} \hat{y}^{(n-j)}\right) + \hat{B}_0^{(n)} \hat{u}^{(n)} \tag{22}$$

$$\hat{y}^{(n)} = \hat{C}^{(n)} \hat{x}^{(n)} \tag{23}$$

$$\hat{u}^{(n)} = \sum_{j=0}^{r} \left(\hat{K}_j^{(n)} \hat{y}^{(n-j)} + \hat{K}_j^{a(n)} y^{(n-j)}\right) = \sum_{j=0}^{r} \left(\hat{K}_j^{(n)} \hat{C}^{(n)} \hat{x}^{(n-j)} + \hat{K}_j^{a(n)} C^{(n)} x^{(n-j)}\right) \tag{24}$$

$\forall n \in \mathbf{Z}_{0+}$, where $\hat{x}^{(n+1)} \in \mathbf{R}^{\eta_{n+1}}$ is the "a posteriori" state of the aggregated subsystem at the $(n+1)$-th iteration whose "a priori" value is $\hat{x}^{(n)} \in \mathbf{R}^{\eta_n}$, $\hat{y}^{(n)} \in \mathbf{R}^{p_n}$, $\hat{u}^{(n)} \in \mathbf{R}^{m_n}$, $\hat{A}^{(n+1)} \in \mathbf{R}^{\eta_{n+1} \times (\sum_{i=0}^{n} n_i)}$, $\hat{B}_0^{(n)} \in \mathbf{R}^{\eta_{n+1} \times m_n}$, $\hat{B}_j^{(n)} \in \mathbf{R}^{\eta_{n+1} \times p_{n-j}}$, $\hat{B}_j^{a(n)} \in \mathbf{R}^{\eta_{n+1} \times (\sum_{i=0}^{n-j} p_i)}$ for $j \in \bar{r}$; $\hat{C}^{(n-j)} \in \mathbf{R}^{p_{n-j} \times n_{n-j}}$, $\hat{K}_j^{(n)} \in \mathbf{R}^{m_{n+1} \times p_{n-j}}$, $\hat{K}_j^{a(n)} \in \mathbf{R}^{m_{n+1} \times (\sum_{i=0}^{n-j} p_i)}$ for $j \in \bar{r} \cup \{0\}$, and $\hat{D}^{(n)}$, $\widetilde{\hat{D}}^{(n)} \in \mathbf{R}^{\eta_{n+1} \times n_n}$ and $\hat{B}_0^{(n)} \in \mathbf{R}^{\eta_{n+1} \times m_{n+1}}$, $\hat{B}_j^{(n)} \in \mathbf{R}^{\eta_{n+1} \times p_{n-j}}$ for $j \in \bar{r}$; $\forall n \in \mathbf{Z}_{0+}$.

Note that the aggregated subsystem (22)–(24) is coupled to the former global state $x^{(n)}$ describing the total system's dynamics prior to the aggregation action. It can be seen that the coupling terms do not necessary demonstrate infinite memory requirements as n tends to infinity, since the matrices $\hat{A}^{(n+1)}$, $\hat{B}_j^{a(n)}$ and $\hat{K}_j^{a(n)}$ can contain nonzero columns associated with the most recent state/output data related to the previous aggregation system; see, for instance, [12]. Note also that, due to the coupling between the a priori whole state at the n-th iterations with the a priori new aggregated substate, it can happen that the a posteriori vector after the new aggregated substate has a higher dimension than its a priori version. The various dynamics, control and output matrices have the appropriate orders.

After incorporating the control law, we can write this whole system of extended states $x^{(n)} = \left(x^{0(n)^T}, \hat{x}^{(n)^T}\right)^T \in \mathbf{R}^{\sum_{i=0}^{n} n_i}$; $\forall n \in \mathbf{Z}_{0+}$ in a compact way:

$$x^{(n+1)} = \begin{bmatrix} x^{0(n+1)} \\ \hat{x}^{(n+1)} \end{bmatrix} = \begin{bmatrix} A^{0(n)} + B_0^{(n)} K_0^{(n)} C^{(n)} & \hat{A}^{0(n)} \\ \hat{A}^{(n+1)} + \hat{B}_0^{(n)} \hat{K}_0^{a(n)} C^{(n)} & \hat{D}^{(n)} + \widetilde{\hat{D}}^{(n)} + \hat{B}_0^{(n)} \hat{K}_0^{(n)} \hat{C}^{(n)} \end{bmatrix} \begin{bmatrix} x^{(n)} \\ \hat{x}^{(n)} \end{bmatrix}$$

$$+ \sum_{j=1}^{r} \begin{bmatrix} \left(B_j^{(n)} + B_0^{(n)} K_j^{(n)}\right) C^{(n-j)} & 0 \\ \left(\hat{B}_j^{a(n)} + \hat{B}_0^{(n)} \hat{K}_j^{a(n)}\right) C^{(n-j)} & \left(\hat{B}_j^{(n)} + \hat{B}_0^{(n)} \hat{K}_j^{(n)}\right) \hat{C}^{(n-j)} \end{bmatrix} x^{(n-j)}; \forall n \in \mathbf{Z}_{0+} \qquad (25)$$

so that $x^{(n)}, x^{0(n+1)} \in \mathbf{R}^{\sum_{i=0}^{n} n_i}$ and $\hat{x}^{(n)} \in \mathbf{R}^{\eta_n}$ and $\hat{x}^{(n+1)} \in \mathbf{R}^{\eta_{n+1}}$ imply that $x^{(n+1)} \in \mathbf{R}^{\sum_{i=0}^{n+1} n_i}$, $x^{(n+1)} = \left(x^{0(n+1)^T}, \hat{x}^{(n+1)^T}\right)^T \in \mathbf{R}^{\sum_{i=0}^{n+1} n_i}$.

In order to construct a state vector which includes delayed dynamics, we now define the modified extended state $\bar{x}^{(n)}$ defined by $\bar{x}^{(n)} = \left(x^{(n)^T}, x^{(n-1)^T}, \ldots, x^{(n-r)^T}\right)^T \in \mathbf{R}^{\sum_{j=n-r}^{n} \sum_{i=0}^{j} n_i}$; $\forall n \in \mathbf{Z}_{0+}$. Thus, one determines from (25) that:

$$\bar{x}^{(n+1)} = \overline{A}^{(n)} \bar{x}^{(n)}; \forall n \in \mathbf{Z}_{0+} \qquad (26)$$

where

$$\overline{A}^{(n)} = \begin{bmatrix} A^{0(n)} + B_0^{(n)} K_0^{(n)} C^{(n)} & \hat{A}^{0(n)} & \left(B_1^{(n)} + B_0^{(n)} K_1^{(n)}\right) C^{(n-1)} & 0 & \cdots & \left(B_r^{(n)} + B_0^{(n)} K_r^{(n)}\right) C^{(n-r)} & 0 \\ \hat{A}^{(n+1)} + \hat{B}_0^{(n)} \hat{K}_0^{a(n)} C^{(n)} & \hat{D}^{(n)} + \widetilde{\hat{D}}^{(n)} + \hat{B}_0^{(n)} \hat{K}_0^{(n)} \hat{C}^{(n)} & \left(\hat{B}_1^{a(n)} + \hat{B}_0^{(n)} \hat{K}_1^{a(n)}\right) C^{(n-1)} & \left(\hat{B}_1^{(n)} + \hat{B}_0^{(n)} \hat{K}_1^{(n)}\right) \hat{C}^{(n-1)} & \cdots & \left(\hat{B}_r^{a(n)} + \hat{B}_0^{(n)} \hat{K}_r^{a(n)}\right) C^{(n-r)} & \left(\hat{B}_r^{(n)} + \hat{B}_0^{(n)} \hat{K}_r^{(n)}\right) \hat{C}^{(n-r)} \\ I_{\sum_{i=0}^{n-1} n_i} & & & \ddots & & & \\ & & I_{\sum_{i=0}^{n-r} n_i} & & \cdots & 0 & 0 \end{bmatrix} \qquad (27)$$

$\forall n \in \mathbf{Z}_{0+}$, with $\overline{A}^{(n)} \in \mathbf{R}^{(\sum_{j=n-r}^{n+1} \sum_{i=0}^{j} n_i) \times (\sum_{j=n-r}^{n} \sum_{i=0}^{j} n_i)}$. Now, consider the symmetric matrices:

$$\overline{M}^{(n)} = \overline{A}^{(n)^T} \overline{A}^{(n)} = \begin{bmatrix} \overline{A}^{(n)^T} \overline{A}^{(n)} & \overline{A}^{(n)^T} \overline{B}^{(n)} \\ \overline{B}^{(n)^T} \overline{A}^{(n)} & \overline{B}^{(n)^T} \overline{B}^{(n)} \end{bmatrix}$$

$$M^{(n)} = \begin{bmatrix} A^{0(n+1)^T} A^{0(n+1)} & A^{0(n+1)^T} B^{0(n+1)} \\ B^{0(n+1)^T} A^{0(n+1)} & B^{0(n+1)^T} B^{0(n+1)} \end{bmatrix} \qquad (28)$$

$\in \mathbf{R}^{(\sum_{i=0}^{n+1} n_i + \sum_{j=n-r}^{n}(\sum_{i=0}^{j} n_i + n_j)) \times (\sum_{i=0}^{n+1} n_i + \sum_{j=n-r}^{n}(\sum_{i=0}^{j} n_i + n_j))}$; $\forall n \in \mathbf{Z}_{0+}$

where the relations between the a priori dynamics of the new iteration after the aggregation of a new substate to the whole dynamics with the a posteriori dynamics of the former iteration are given by:

$$A^{0(n+1)} = \begin{bmatrix} A^{(n)} & 0_{(\sum_{i=0}^{n+1} n_i) \times n_{n+1}} \end{bmatrix}$$

$$= \begin{bmatrix} A^{0(n)} + B_0^{(n)} K_0^{(n)} C^{(n)} & \hat{A}^{0(n)} & 0 \\ \hat{A}^{(n+1)} + \hat{B}_0^{(n)} \hat{K}_0^{a(n)} C^{(n)} & \hat{D}^{(n)} + \widetilde{D}^{(n)} + \hat{B}_0^{(n)} \hat{K}_0^{(n)} \hat{C}^{(n)} & 0 \end{bmatrix} \in R^{(\sum_{i=0}^{n+1} n_i) \times (\sum_{i=0}^{n+1} n_i)} \quad (29)$$

$$B^{0(n+1)} = B^{(n)} = \begin{bmatrix} \left(B_1^{(n)} + B_0^{(n)} K_1^{(n)}\right) C^{(n-1)} & 0 & \cdots & \left(B_j^{(n)} + B_0^{(n)} K_r^{(n)}\right) C^{(n-r)} & 0 \\ \left(\hat{B}_1^{a(n)} + \hat{B}_0^{(n)} \hat{K}_1^{a(n)}\right) C^{(n-1)} & \left(\hat{B}_1^{(n)} + \hat{B}_0^{(n)} \hat{K}_1^{(n)}\right) \hat{C}^{(n-1)} & \cdots & \left(\hat{B}_j^{a(n)} + \hat{B}_0^{(n)} \hat{K}_r^{a(n)}\right) C^{(n-r)} & \left(\hat{B}_j^{(n)} + \hat{B}_0^{(n)} \hat{K}_r^{(n)}\right) \hat{C}^{(n-r)} \end{bmatrix} \quad (30)$$

$$\in R^{(\sum_{i=0}^{n+1} n_i) \times (\sum_{j=n-r}^{n-1} (\sum_{i=0}^{j} n_i + n_j))}$$

$\forall n \in Z_{0+}$. which are built in order to complete a square a priori matrix of dynamics of the $(n+1)$- the aggregated system which was obtained after the aggregation of the $(n+1)$-th subsystem.

The stability of the aggregation dynamic system (19) to (24) under discrete delays is now discussed via the modified extended system (26), subject to (27), which can be obtained via Lemmas 7,8 from the convergence of the symmetric matrix (28), subject to (29),(30). The following result holds:

Theorem 1. *The following properties hold:*

(i) $\{M^{(n)}\}_{n=0}^{\infty}$ and $\{\overline{A}^{(n)}\}_{n=0}^{\infty}$ are convergent, and also asymptotically convergent, if and only if $\lim_{m \to \infty} \overline{A}^{(n)^m} = 0$; $\forall n \in Z_{0+}$.

(ii) $\{M^{(n)}\}_{n=0}^{\infty}$ and $\{\overline{A}^{(n)}\}_{n=0}^{\infty}$ are asymptotically convergent if and only if $\lim_{m \to \infty} \left\{ \left| \|\overline{A}^{(n+\xi)}\| - \|\overline{A}^{(n)}\| \right|^m \right\}_{n=0}^{\infty} = 0$ for any given $n \in Z_{0+}$, $\xi \in Z_+$.

(iii) If $\{\overline{A}^{(n)}\}_{n=0}^{\infty}$ (and then $\{M^{(n)}\}_{n=0}^{\infty}$) is convergent, then the state of the modified extended system, (26), converges asymptotically to zero, i.e. $\overline{x}^{(n+m)} \to 0$ and also $x^{(n+m)} \to 0$ as $m \to \infty$ for any given initial condition $x^{(0)}$ and any $n \in Z_{0+}$ so that the aggregation system is globally asymptotically stable.

(iv) If $\{\overline{A}^{(n)}\}_{n=0}^{\infty}$ (and then $\{M^{(n)}\}_{n=0}^{\infty}$) is asymptotically convergent then $\|\overline{x}^{(n+m+\xi)} - \overline{x}^{(n+m)}\| \to 0$ as $m \to \infty$ for any given $n \in Z_{0+}$, $\xi \in Z_+$ and any given initial condition $x^{(0)}$ and also $\|x^{(n+m+\xi)} - x^{(n+m)}\| \to 0$ as $m \to \infty$ for any given initial condition $x^{(0)}$ and any given $n \in Z_{0+}$, $\xi \in Z_+$ so that the incremental aggregation system is globally asymptotically stable.

(v) Assume that $M^{(0)}$ is convergent and that $\left| \lambda_{max}\left(B^{(n)^T} B^{(n)}\right) \right| < \min\left(1 - \varepsilon^{(n+1)}, \varepsilon^{(n)} - \varepsilon^{(n+1)}\right)$; $\forall n \in Z_{0+}$ for some strictly decreasing real sequence $\{\varepsilon^{(n)}\}_{n=0}^{\infty} \subset [0, 1)$. Then, $\|M^{(n)}\|_2 < 1$; $\forall n \in Z_{0+}$ and $\{M^{(n)}\}_{n=0}^{\infty}$ and $\{A^{(n)}\}_{n=0}^{\infty}$ are convergent sequences.

It is of interest to now discuss how the stability properties of the aggregation system of Theorem 1 can be guaranteed or addressed by the synthesis of the basic controller (21) on the current aggregated system, and how its updated rule (24) can be applied to the new aggregated subsystem to generate the aggregated system for the next iteration step. This discussion invokes conditions to guarantee that the equation of dimensionally compatible real matrices

$$BKC = A_m - A \quad (31)$$

is solvable in K for a given quadruple (A, B, C, A_m) with A and A_m being square, A_m being convergent (basically stable in the discrete context) and defining the closed-loop system dynamics after linear output-feedback control $u = Ky = KCx$ via the linear stabilizing controller of gain K; A, B and C are the open-loop dynamics (i.e., the one being got for $K = 0$) and B and C are the control and output matrices. Equation (31) is written in equivalent vector form for the unknown K as follows:

$$\left(B \otimes C^T\right) vec\left(K\right) = vec(A_m - A) \tag{32}$$

It turns out that (31) is solvable in K if and only if (32) is solvable in $vec\left(K\right)$, that is, if $rank\left(B \otimes C^T\right) = rank\left[B \otimes C^T, vec(A_m - A)\right]$ according to the Rouché-Froebenius theorem for solvability of linear systems of algebraic equations. Note that if A_m satisfies the constraint $A_m = EA$, for some square matrix E of the same order as A, then $\|A_m\|_2 < 1$ (so that A_m is convergent) if $\|E\|_2 < 1/\|A\|_2$. In particular, if $A_m = \rho A$ with $\rho \in \mathbf{R}$ then A_m is convergent if $|\rho| < 1/\|A\|_2$. A preliminary technical result concerning the solvability if the concerned algebraic system (31), or equivalently (32), is (either indeterminate or determinate) compatible to be then used follows:

Lemma 9. *Assume that $A \in \mathbf{R}^{n \times n}$, $B \in \mathbf{R}^{m \times n}$ and $C \in \mathbf{R}^{p \times n}$. Then, the following properties hold:*

(i) *linear output-feedback controller exists which stabilizes the closed-loop matrix of dynamics $A_m = EA$ for some E with $\|E\|_2 < 1/\|A\|_2$, [5,7], which satisfies the rank constraint:*

$$rank\left(B \otimes C^T\right) = rank\left[B \otimes C^T, \left(I \otimes A^T\right) vec(E) - vec(A)\right] \tag{33}$$

If (33) holds, then the set of stabilizing linear-output feedback controllers of gains K which solve (32), equivalently (31), which is a compatible algebraic linear system, for $A_m = EA$, are given by

$$vec\left(K\right) = \left(B \otimes C^T\right)^\dagger \left[\left(I \otimes A^T\right) vec(E) - vec(A)\right] + \left[I - \left(B \otimes C^T\right)^\dagger \left(B \otimes C^T\right)\right] k_w \tag{34}$$

with k_w being any arbitrary real vector of the same dimension as $vec\left(K\right)$. Assume that $\left(B \otimes C^T\right)^\dagger = \left(B \otimes C^T\right)^{-1}$ (a necessary condition being $\min(m, p) \geq n$). Then (32) for $A_m = EA$ is a compatible determinate, and the unique solution to (33) is

$$vec\left(K\right) = \left(B \otimes C^T\right)^\dagger \left[\left(I \otimes A^T\right) vec(E) - vec(A)\right] \tag{35}$$

If $A_m = \rho A$ with $|\rho| < 1/\|A\|_2$ then (33), (34) and (35) become, in particular,

$$rank\left(B \otimes C^T\right) = rank\left[B \otimes C^T, (\rho - 1) vec(A)\right] \tag{36}$$

$$vec\left(K\right) = (\rho - 1)\left(B \otimes C^T\right)^\dagger vec(A) + \left[I - \left(B \otimes C^T\right)^\dagger \left(B \otimes C^T\right)\right] k_w \tag{37}$$

and

$$vec\left(K\right) = \left(B \otimes C^T\right)^\dagger \left(B \otimes C^T\right)^\dagger vec(A) \tag{38}$$

(ii) *Assume that $A_m = EA$ and*

$$rank\left[B \otimes C^T, \left(I \otimes A^T\right) vec(E) - vec(A)\right] = rank\left(B \otimes C^T\right) + 1 \tag{39}$$

Then, (32), equivalently (31), is an algebraically incompatible system of equations, and

$$vec\left(K\right) = \left(B \otimes C^T\right)^\dagger \left[\left(I \otimes A^T\right) vec(E) - vec(A)\right], \text{ i.e.} \tag{40}$$

i.e., Equation (34) for $k_w = 0$, is the best least-squares approximated solution to (32) in the sense that the corresponding controller gain minimizes the norm error $\|BKC + A - A_m\|_2^2$. If (39) holds for any A_m of the form $A_m = EA$ then there is no solution to (31) in K; only best approximation solutions exist.

Particular cases of interest which are well-known from basic Control Theory (see e.g., [13]) are:

(1) $p = n$, $C \in \mathbf{R}^{n \times n}$ is non-singular and (A, B) is stabilizable, i.e., any unstable or critically unstable mode of the open-loop dynamics can be closed-loop stabilized under linear state feed-back control. Thus, $\mathrm{rank}\,[\lambda I_n - A, B] = n$; $\forall \lambda \in \mathbf{C}$ with $|\lambda| \geq 1$ (discrete form of Popov-Belevitch-Hautus stabilizability test [6,13,14]). Then (31) becomes $BK = (A_m - A)C^{-1}$, which is solvable in K, and there is always an output-feedback stabilizing linear controller generating a stabilizing controller of gain K, generating a control $u = Ky = KCx$, such that the closed-loop dynamics is defined by a convergent matrix $A_m = A + BKC$.

(2) In Case 1, $C = I_n$. Then, the control law is a linear state-feedback control, and a state-feedback stabilizing linear controller generating a control $u = Kx$ exists, leading to closed-loop dynamics defined by the convergent matrix $A_m = A + BK$.

Lemma 9 is useful to guarantee the relevant results of Theorem 1 in terms of the controller gains choices under certain algebraic solvability conditions. This feature is addressed in the subsequent result:

Theorem 2. *Assume that:*

(1) $\mathrm{rank}\left(B_0^{(n)} \otimes C^{(n)T}\right) = \mathrm{rank}\left(B_0^{(n)} \otimes C^{(n)T}, A_f^{(n)} - A^{0(n)}\right)$ so that $A^{0(n)} + B_0^{(n)} K_0^{(n)} C^{(n)} = A_f^{(n)}$ is solvable in $K_0^{(n)}$ for some convergent matrix $A_f^{(n)}$ of appropriate order; $\forall n \in \mathbf{Z}_{0+}$, $\mathrm{rank}\left(\hat{B}_0^{(n)} \otimes \hat{C}^{(n)T}\right) = \mathrm{rank}\left(\hat{B}_0^{(n)} \otimes \hat{C}^{(n)T}, \hat{A}_f^{a(n+1)} - \hat{A}^{(n+1)}\right)$ so that $\hat{A}^{(n+1)} + \hat{B}_0^{(n)} \hat{K}_0^{(n)} \hat{C}^{(n)} = \hat{A}_f^{a(n+1)}$ is solvable in $\hat{K}_0^{a(n)}$ for some matrix $\hat{A}_f^{a(n+1)}$ of appropriate order; $\forall n \in \mathbf{Z}_{0+}$, $\mathrm{rank}\left(\hat{B}_0^{(n)} \otimes \hat{C}^{(n)T}\right) = \mathrm{rank}\left(\hat{B}_0^{(n)} \otimes \hat{C}^{(n)T}, \hat{A}_f^{(n+1)} - \hat{D}^{(n)} - \widetilde{D}^{(n)}\right)$ so that $\hat{A}_f^{(n)} = \hat{D}^{(n)} + \widetilde{D}^{(n)} + \hat{B}_0^{(n)} \hat{K}_0^{(n)} \hat{C}^{(n)}$ is solvable in $\hat{K}_0^{(n)}$ for some matrix $\hat{A}_f^{(n)}$ of appropriate order; $\forall n \in \mathbf{Z}_{0+}$,

(2) *and that subsequent rank conditions hold:*

$$\mathrm{rank}\left(B_0^{(n)} \otimes C^{(n-i)T}\right) = \mathrm{rank}\left(B_0^{(n)} \otimes C^{(n-i)T}, B_i^{(n)} C^{(n-i)}\right)$$
$$\mathrm{rank}\left(\hat{B}_0^{(n)} \otimes \hat{C}^{(n-i)T}\right) = \mathrm{rank}\left(\hat{B}_0^{(n)} \otimes \hat{C}^{(n-i)T}, \hat{B}_i^{(n)} \hat{C}^{(n-i)}\right)$$
$$\mathrm{rank}\left(\hat{B}_0^{(n)} \otimes C^{(n-i)T}\right) = \mathrm{rank}\left(\hat{B}_0^{(n)} \otimes C^{(n-i)T}, \hat{B}_i^{a(n)} C^{(n-i)}\right)$$

$\forall i \in \overline{r}$ so that the following matrix equations are solvable in the delayed controller gains $K_i^{(n)}$, $\hat{K}_i^{(n)}$ and $\hat{K}_i^{a(n)}$:

$$B_0^{(n)} K_i^{(n)} C^{(n-i)} = -B_i^{(n)} C^{(n-i)}; \hat{B}_0^{(n)} \hat{K}_i^{(n)} \hat{C}^{(n-i)} = -\hat{B}_i^{(n)} \hat{C}^{(n-i)};$$
$$\hat{B}_0^{(n)} \hat{K}_i^{a(n)} C^{(n-i)} = -\hat{B}_i^{(n)} C^{(n-i)}; \forall i \in \overline{r}; \forall n \in \mathbf{Z}_{0+}.$$

Then, the matrix equations

$$A^{0(n)} + B_0^{(n)} K_0^{(n)} C^{(n)} = A_f^{(n)}, \; \hat{A}^{(n+1)} + \hat{B}_0^{(n)} \hat{K}_0^{a(n)} C^{(n)} = \hat{A}_f^{a(n+1)},$$
$$\hat{A}_f^{(n)} = \hat{D}^{(n)} + \widetilde{D}^{(n)} + \hat{B}_0^{(n)} \hat{K}_0^{(n)} \hat{C}^{(n)}, \; B_0^{(n)} K_i^{(n)} C^{(n-i)} = -B_i^{(n)} C^{(n-i)}; \quad (41)$$
$$\hat{B}_0^{(n)} \hat{K}_i^{(n)} \hat{C}^{(n-i)} = -\hat{B}_i^{(n)} \hat{C}^{(n-i)}; \; \forall i \in \overline{r}; \; \forall n \in \mathbf{Z}_{0+})$$

are solvable in the controller gains $K_0^{(n)}$, $K_i^{(n)}$ and $\hat{K}_i^{(n)}$; $\forall i \in \bar{r}$; $\forall n \in Z_{0+}$ leading to the solutions

$$vec\left(K_0^{(n)}\right) = \left(B_0^{(n)} \otimes C^{(n)^T}\right)^{\dagger} vec\left(A_f^{(n)} - A^{0(n)}\right) + \left(I - \left(B_0^{(n)} \otimes C^{(n)^T}\right)^{\dagger}\left(B_0^{(n)} \otimes C^{(n)^T}\right)\right) vec\left(K_0^{v(n)}\right)$$

$$vec\left(\hat{K}_0^{a(n)}\right) = \left(\hat{B}_0^{(n)} \otimes C^{(n)^T}\right)^{\dagger} vec\left(\hat{A}_f^{a(n+1)} - \hat{A}^{(n+1)}\right) + \left(I - \left(\hat{B}_0^{(n)} \otimes C^{(n)^T}\right)^{\dagger}\left(\hat{B}_0^{(n)} \otimes C^{(n)^T}\right)\right) vec\left(\hat{K}_0^{av(n)}\right)$$

$$vec\left(\hat{K}_0^{(n)}\right) = \left(\hat{B}_0^{(n)} \otimes \hat{C}^{(n)^T}\right)^{\dagger} vec\left(\hat{A}_f^{(n+1)} - \hat{D}^{(n)} - \tilde{D}^{(n)}\right) + \left(I - \left(\hat{B}_0^{(n)} \otimes \hat{C}^{(n)^T}\right)^{\dagger}\left(\hat{B}_0^{(n)} \otimes \hat{C}^{(n)^T}\right)\right) vec\left(\hat{K}_0^{v(n)}\right) \quad (42)$$

$$vec\left(K_i^{(n)}\right) = -\left(B_0^{(n)} \otimes C^{(n-i)^T}\right)^{\dagger} vec\left(B_i^{(n)} C^{(n-i)}\right) + \left(I - \left(B_0^{(n)} \otimes C^{(n-i)^T}\right)^{\dagger}\left(B_0^{(n)} \otimes C^{(n-i)^T}\right)\right) vec\left(K_i^{v(n)}\right)$$

$$vec\left(\hat{K}_i^{(n)}\right) = -\left(\hat{B}_0^{(n)} \otimes \hat{C}^{(n-i)^T}\right)^{\dagger} vec\left(\hat{B}_i^{(n)} \hat{C}^{(n-i)}\right) + \left(I - \left(\hat{B}_0^{(n)} \otimes \hat{C}^{(n-i)^T}\right)^{\dagger}\left(\hat{B}_0^{(n)} \otimes \hat{C}^{(n-i)^T}\right)\right) vec\left(\hat{K}_i^{v(n)}\right)$$

$$vec\left(\hat{K}_i^{a(n)}\right) = -\left(\hat{B}_0^{(n)} \otimes C^{(n-i)^T}\right)^{\dagger} vec\left(\hat{B}_i^{a(n)} C^{(n-i)}\right) + \left(I - \left(\hat{B}_0^{(n)} \otimes C^{(n-i)^T}\right)^{\dagger}\left(\hat{B}_0^{(n)} \otimes C^{(n-i)^T}\right)\right) vec\left(\hat{K}_i^{av(n)}\right)$$

with $K_0^{v(n)}$, $K_0^{av(n)}$, $\hat{K}_0^{v(n)}$, $K_i^{v(n)}$, $\hat{K}_i^{v(n)}$ and $\hat{K}_i^{av(n)}$; $\forall i \in \bar{r}$; $\forall n \in Z_{0+}$ being arbitrary matrices of appropriate orders for the corresponding equation (above) in each case whose equivalent vector expressions are denoted by $vec(.)$.

It should be pointed out that it can be of interest to apply the results on interlacing Cauchy's theorem and some of its extensions (see e.g., [20–22]) to the stability of aggregation models based on dynamic systems formulated via differential, difference or hybrid differential/difference equations.

6. Conclusions

This paper relies on partitioned Hermitian matrices and Cauchy's interlacing theorem and the associated stability results. Based on the fact that convergent matrices are a discrete counterpart of stability matrices, the results presented above are then extended to sequences of convergent matrices. Then, an example of a SIR-type epidemic model continuous-time consisting of intercommunity clusters is discussed relative to the previously given stability theoretic results under the proposed framework based on Cauchy's interlacing theorem, and which may be of interest for healthcare management. Later, a dynamic linear discrete aggregation model is discussed, which involves output delay and linear output feedback, and which can be also be reformulated for linear-state feedback by identifying state and output, that is, by taking the output matrix equal to the identity, provided that the state is available for measurement. The studied aggregation model is built through the successive incorporation of discrete subsystems with particular coupled dynamics. Stabilizing decentralized controllers are proposed and discussed for this type of aggregation model.

Author Contributions: The author contributed by himself the whole manuscript.

Funding: This research has been jointly supported by the Spanish Government and the European Commission with Grants RTI2018-094336-B-I00 (MINECO/FEDER, UE) and DPI2015-64766-R (MINECO/FEDER, UE).

Acknowledgments: The author is grateful to the Spanish Government for Grants RTI2018-094336-B-I00 and DPI2015-64766-R (MINECO/FEDER, UE).

Conflicts of Interest: The author declares that he has no competing interests regarding the publication of this article.

Appendix A Mathematical Proofs

Proof of Lemma 1. From the given assumptions, M is Hermitian, and note that

$$\det M\prime = \det\begin{bmatrix} M & 0 \\ m^* & d \end{bmatrix} + \det\begin{bmatrix} M & m \\ m^* & \tilde{d} \end{bmatrix} = \det\begin{bmatrix} M & 0 \\ m^* & \tilde{d} \end{bmatrix} + \det\begin{bmatrix} M & m \\ m^* & d \end{bmatrix} = \det\begin{bmatrix} M & 0 \\ m^* & d+\tilde{d} \end{bmatrix} + \det\begin{bmatrix} M & m \\ m^* & 0 \end{bmatrix}.$$

Thus, Properties (i) to (iii) follow directly from the above relations by noting, furthermore, that $\det\begin{bmatrix} M & 0 \\ m^* & c \end{bmatrix} = c \det M$. □

Proof of Lemma 2. First, note that

$$\left| k_d^{(n)} - k_{\widetilde{d}}^{(n)} \right| \leq \|d^{(n)}| - |\widetilde{d}^{(n)}\| \leq |d^{(n)} + \widetilde{d}^{(n)}| \leq K_d^{(n)} + K_{\widetilde{d}}^{(n)}; \quad \forall n (\geq n_0) \in \mathbb{Z}_+.$$

Thus,

$$\det M^{(n+1)} = \det\begin{bmatrix} M^{(n)} & 0 \\ m^{(n)*} & d^{(n)} \end{bmatrix} + \det\begin{bmatrix} M^{(n)} & m^{(n)} \\ m^{(n)*} & \widetilde{d}^{(n)} \end{bmatrix}$$

$$= d^{(n)} \det M^{(n)} + \det\left(\begin{bmatrix} M^{(n)} & 0 \\ 0 & \widetilde{d}^{(n)} \end{bmatrix}\left(I_{n+1} + \begin{bmatrix} M^{(n)-1} & 0 \\ 0 & 1/\widetilde{d}^{(n)} \end{bmatrix}\begin{bmatrix} 0 & m^{(n)} \\ m^{(n)*} & 0 \end{bmatrix}\right)\right) \quad (A1)$$

$$= d^{(n)} \det M^{(n)} + \widetilde{d}^{(n)} \det M^{(n)} \det\left(I_{n+1} + \begin{bmatrix} M^{(n)-1} & 0 \\ 0 & 1/\widetilde{d}^{(n)} \end{bmatrix}\begin{bmatrix} 0 & m^{(n)} \\ m^{(n)*} & 0 \end{bmatrix}\right);$$
$$\forall n (\geq n_0) \in \mathbb{Z}_+$$

if $\left\{ M^{(n)-1} \right\}_{n=n_0}^{\infty}$ exists so that $\det M^{(n)} \neq 0$, that is, if $0 < k_M^{(n)} = \left|\lambda_{\min}\left(M^{(n)}\right)\right| \leq K_M^{(n)} = \left|\lambda_{\max}\left(M^{(n)}\right)\right| < +\infty$; $\forall n(\geq n_0) \in \mathbb{Z}_+$. Thus,

$$\left|\det M^{(n+1)}\right| \leq \left|d^{(n)} \det M^{(n)}\right| + \left|\widetilde{d}^{(n)} \det M^{(n)}\right| \left|\det\left(I_{n+1} + \begin{bmatrix} M^{(n)-1} & 0 \\ 0 & 1/\widetilde{d}^{(n)} \end{bmatrix}\begin{bmatrix} 0 & m^{(n)} \\ m^{(n)*} & 0 \end{bmatrix}\right)\right|$$

$$\leq n K_d^{(n)} K_M^{(n)} + n K_{\widetilde{d}}^{(n)} K_M^{(n)} \left(1 + \frac{1}{k_M^{(n)} k_{\widetilde{d}}^{(n)}} \sqrt{m^{(n)*} m^{(n)}}\right)^{n+1} \quad (A2)$$

$$\leq n K_d^{(n)} K_M^{(n)} + n K_{\widetilde{d}}^{(n)} K_M^{(n)} \left(1 + \varepsilon^{(n)}\right)^{n+1} \leq n K_d^{(n)} K_M^{(n)} + n K_{\widetilde{d}}^{(n)} K_M^{(n)} \left(1 + (2^{n+1} - 1)\varepsilon^{(n)}\right);$$
$$\forall n(\geq n_0) \in \mathbb{Z}_+$$

since $\lambda_{\max}(m^{(n)*} m^{(n)}) = m^{(n)*} m^{(n)}$; $\forall n(\geq n_0) \in \mathbb{Z}_+$. Since it is assumed the existence of a real sequence $\{\varepsilon^{(n)}\}_{n=n_0}^{\infty} \subset [0,1)$ such that $\frac{1}{k_M^{(n)} k_{\widetilde{d}}^{(n)}} \sqrt{m^{(n)*} m^{(n)}} \leq \varepsilon^{(n)}$; $\forall n(\geq n_0) \in \mathbb{Z}_+$ and since

$$\left(1 + \varepsilon^{(n)}\right)^{n+1} = \sum_{i=0}^{n+1} \binom{n+1}{i} \varepsilon^{(n)i} = 1 + \sum_{i=1}^{n+1} \binom{n+1}{i} \varepsilon^{(n)i} \leq 1 + (2^{n+1} - 1)\varepsilon^{(n)}; \quad (A3)$$
$$\forall n(\geq n_0) \in \mathbb{Z}_+$$

thus, it follows that $\left|\det M^{(n+1)}\right| \leq (n+1) K_M^{(n+1)} \leq n K_M^{(n)}$ holds for any given $n(\geq n_0) \in \mathbb{Z}_+$ if

$$n K_d^{(n)} K_M^{(n)} + n K_{\widetilde{d}}^{(n)} K_M^{(n)} \left(1 + (2^{n+1} - 1)\varepsilon^{(n)}\right) \leq n K_M^{(n)}$$

for any given $n(\geq n_0) \in \mathbb{Z}_+$, that is, if

$$n K_M^{(n)} \left(K_d^{(n)} - 1\right) + n K_{\widetilde{d}}^{(n)} K_M^{(n)} \left(1 + (2^{n+1} - 1)\varepsilon^{(n)}\right) \leq 0$$

for such $n \in \mathbb{Z}_+$, equivalently, if $nK_d^{(n)}\left(1 + (2^{n+1} - 1)\varepsilon^{(n)}\right) \leq n\left(1 - K_d^{(n)}\right)$, equivalently if $K_d^{(n)} \leq 1$, and furthermore, and accordingly to the restriction on the sequence $\{\varepsilon^{(n)}\}_{n=n_0}^{\infty}$, if

$$K_d^{(n)} \leq \frac{\left(1 - K_d^{(n)}\right)}{1 + (2^{n+1} - 1)\varepsilon^{(n)}} \leq \frac{\left(1 - K_d^{(n)}\right)k_M^{(n)}k_d^{(n)}}{k_M^{(n)}k_d^{(n)} + (2^{n+1} - 1)\sqrt{\lambda_{\max}\left(m^{(n)*}m^{(n)}\right)}} \quad (A4)$$

for such an $n \in \mathbb{Z}_{0+}$. In particular, if $|d^{(n)}| = K_d^{(n)} = k_d^{(n)} \leq 1$ and $|\widetilde{d}^{(n)}| = K_d^{(n)} = k_d^{(n)}$ for some given $n (\geq n_0) \in \mathbb{Z}_{0+}$, then (A4) holds if

$$|\widetilde{d}^{(n)}| \leq \frac{\left(1 - |d^{(n)}|\right)}{1 + (2^{n+1} - 1)\varepsilon^{(n)}} \leq \frac{\left(1 - |d^{(n)}|\right)k_M^{(n)}|d^{(n)}|}{k_M^{(n)}|d^{(n)}| + (2^{n+1} - 1)\sqrt{\lambda_{\max}\left(m^{(n)*}m^{(n)}\right)}} \quad (A5)$$

and Property (i) has been proved. On the other hand, since Property (ii) assumes that the constraints of Property (i) hold the constraint (1) is a direct consequence of Property (i). Also, if $M^{(n)} \geq 0$ and $M^{(n+1)} \geq 0$ then the constraints (2) follow directly from (1) and Cauchy's interlacing theorem of the eigenvalues which are real non-negative. Property (ii) has been proved. Property (iii) follows since $\limsup_{n\to\infty}\left(\det|M^{(n+1)}| - \det|M^{(n)}|\right) \leq 0$ directly from (A1) and the given assumptions. □

Proof of Lemma 3. By the inversion of a block partitioned matrix, one gets:

$$M^{(n+1)^{-1}} = \begin{bmatrix} M^{(n)} & m^{(n)} \\ m^{(n)*} & d^{(n)} + \widetilde{d}^{(n)} \end{bmatrix}^{-1} = \begin{bmatrix} \overline{M}^{(n)} & \overline{m}^{(n)} \\ \overline{m}^{(n)*} & \overline{d}^{(n)} + \overline{\widetilde{d}}^{(n)} \end{bmatrix} \quad (A6)$$

where

$$\overline{M}^{(n)} = M^{(n)^{-1}}\left(I_n + m^{(n)}\left(d^{(n)} + \widetilde{d}^{(n)} - m^{(n)*}M^{(n)^{-1}}m^{(n)}\right)^{-1}m^{(n)*}M^{(n)^{-1}}\right) \quad (A7)$$

$$\overline{m}^{(n)} = -M^{(n)^{-1}}m^{(n)}\left(d^{(n)} + \widetilde{d}^{(n)} - m^{(n)*}M^{(n)^{-1}}m^{(n)}\right)^{-1} \quad (A8)$$

$$\overline{d}^{(n)} + \overline{\widetilde{d}}^{(n)} = \left(d^{(n)} + \widetilde{d}^{(n)} - m^{(n)*}M^{(n)^{-1}}m^{(n)}\right)^{-1}$$

since $M^{(n)}$ is non-singular and $\widetilde{d}^{(n)} \neq m^{(n)*}M^{(n)^{-1}}m^{(n)} - d^{(n)}$ (i.e., $\left(d^{(n)} + \widetilde{d}^{(n)} - m^{(n)*}M^{(n)^{-1}}m^{(n)}\right)$ is non-singular). The proof follows directly from Lemma 2 [(ii)-(iii)] by replacing $M^{(n+1)} \to \overline{M}^{(n+1)} = \overline{M}^{(n+1)^{-1}}$; $\forall n (\geq n_0) \in \mathbb{Z}_+$. □

Proof of Lemma 4. Since $M^{(n_0)} > 0$ then its inverse is also positive definite and then both of them fulfil the positive semi-definiteness constraint of Lemma 2 [(ii),(iii)] and Lemma 3. The proof follows directly since $\sup\max_{n \geq n_0}\left(|\det M^{(n)}|, |\det M^{(n)^{-1}}|\right) < +\infty$ and $\limsup_{n \to \infty}\left(\sup\max\left(|\det M^{(n)}|, |\det M^{(n)^{-1}}|\right)\right) < +\infty$ then $M^{(n)} > 0$ for $n \geq n_0$ if $M^{(n_0)} > 0$, $\limsup_{n\to\infty} M^{(n)} > 0$ and $\liminf_{n\to\infty} M^{(n)} > 0$. □

Proof of Lemma 5. Note that $M^{(n_0)} = A^{(n_0)*}A^{(n_0)} > 0$ since $A^{(n_0)}$ is a stability matrix. From Lemma 2 and Lemma 3, the sequence $\{M^{(n)}\}_{n=n_0}^{\infty}$ consists of positive definite members. Thus, the singular values of the elements of the sequence $\{A^{(n)}\}_{n=n_0}^{\infty}$ are positive and bounded. Since $A^{(n_0)}$ is stable and since the eigenvalues of any square matrix are continuous functions of its entries there is no zero eigenvalue in any member of the sequence $\{A^{(n)}\}_{n=n_0}^{\infty}$. Any member of this sequence has no eigenvalues at the imaginary complex axis other than zero (i.e., any nonzero critically stable eigenvalues) since then the corresponding $M^{(n)} = A^{(n)*}A^{(n)}$ is not positive definite contradicting the given assumption. As a result, no member of the sequence $\{A^{(n)}\}_{n=n_0}^{\infty}$ has a critically stable eigenvalue (i.e., located on the imaginary complex axis) or unstable eigenvalue (i.e., located on the complex open right half plane).

$M^{(n_0)} > 0$ for some given arbitrary $n_0 \in \mathbf{Z}_+$ and assume also that the conditions of Lemma 2 (iii) and Lemma 3 hold. Then, $M^{(n+1)} > 0$; $\forall n (\geq n_0) \in \mathbf{Z}_+$. □

Proof of Lemma 7. It is direct since $\lim_{m \to \infty} \{A^{(n)m}\}_{n=0}^{\infty} = 0$ implies that $\lim_{m \to \infty} \{\|A^{(n+\xi)}\| - \|A^{(n)}\|\|^m\}_{n=0}^{\infty} = 0$ for any given $\xi \in \mathbf{Z}_+$. □

Proof of Lemma 8. If for any given $n \in \mathbf{Z}_{0+}$, $\|A^{(n)}\|_2 < 1$ then $\lim_{m \to \infty} A^{(n)m} = 0$ for any $n \in \mathbf{Z}_{0+}$, then $\lim_{m \to \infty} \{A^{(n)m}\}_{n=0}^{\infty} = 0$, $\|M^{(n)}\|_2^2 = \|A^{(n)*}A^{(n)}\|_2^2 \leq \|A^{(n)}\|_2^4 < 1$ and then $\|M^{(n)}\|_2 < 1$, $\lim_{m \to \infty} M^{(n)m} = 0$ and $\lim_{m \to \infty} \{M^{(n)m}\}_{n=0}^{\infty} = 0$ for any given $n \in \mathbf{Z}_{0+}$. Thus, if $\{A^{(n)}\}_{n=0}^{\infty}$ is convergent then $\{M^{(n)}\}_{n=0}^{\infty}$ is convergent, hence Property (i) holds. On the other hand, since $M^{(n)}$ is semidefinite positive Hermitian by construction; $\forall n \in \mathbf{Z}_{0+}$ then the condition $\|M^{(n)}\|_2 < 1$; $\forall n \in \mathbf{Z}_{0+}$ leads to the convergence of $\{M^{(n)}\}_{n=1}^{\infty}$, and if $\|M^{(n)}\|_2 < 1$ for some $n \in \mathbf{Z}_{0+}$ then

$$r^2(M^{(n)}) = \lambda_{\max}(M^{(n)*}M^{(n)}) = \lambda_{\max}^2(A^{(n)*}A^{(n)}) = \lambda_{\max}^2(M^{(n)}) = \|M^{(n)}\|_2^2 < 1$$

Then, $\|A^{(n)}\|_2 = \lambda_{\max}^{1/2}(A^{(n)*}A^{(n)}) = \|M^{(n)}\|_2^{1/2} < 1$. Thus, if the above holds for any $n \in \mathbf{Z}_{0+}$, one concludes that $\{A^{(n)}\}_{n=0}^{\infty}$ is convergent if $\{M^{(n)}\}_{n=0}^{\infty}$ is convergent. Hence, Property (ii) follows. Property (iii) is a combination of the other two properties since for any $n \in \mathbf{Z}_{0+}$, $\|M^{(n)}\|_2 < 1$, $\|A^{(n)}\|_2 < 1$ implies that $\|M^{(n)}\|_2 < 1$ and $\|M^{(n)}\|_2 < 1$ implies that $\|A^{(n)}\|_2 < 1$. □

Proof of Theorem 1. Properties (i),(ii) follows from Lemma 7 and Lemma 8 by taking into account the factorization (28). Property (iii) follows from Property (i) and (20) since

$$\|\overline{x}^{(n+m)}\| \leq \left(\prod_{i=0}^{m-1} \|\overline{A}^{(n-i)}\|\right) \|\overline{x}^{(n)}\| \leq \max_{0 \leq i \leq m-1} \|\overline{A}^{(n-i)}\|^m \|\overline{x}^{(n)}\|; \quad \forall n \in \mathbf{Z}_{0+}, \forall m \in \mathbf{Z}_+ \tag{A9}$$

and then $\overline{x}^{(n+m)} \to 0$ as $m \to \infty$ for any $n \in \mathbf{Z}_{0+}$. Property (iv) is proved in the same way as Property (iii) via Property (ii). The proof of Property (v) is made by comparing (28) with (10)–(12) by replacing $\lambda^{(n)}\overline{M}^{(n)} \to A^{(n)^T}B^{(n)}$ and $\Delta^{(n)} \to B^{(n)^T}B^{(n)}$. One gets, via complete induction, that if $M^{(0)}$ is convergent and, furthermore,

$$\lambda_{\max}\left(B^{(n)^T}A^{(n)}A^{(n)^T}B^{(n)}\right) < \left(1 - \varepsilon^{(n+1)} - \max\left(1 - \varepsilon^{(n)}, \left|\lambda_{\max}\left(B^{(n)^T}B^{(n)}\right)\right|\right)\right)^2; \tag{A10}$$
$$\forall n \in \mathbf{Z}_{0+}$$

Then, $\{M^{(n)}\}_{n=0}^{\infty}$ is convergent, since $M^{(0)}$ is convergent, provided that $\varepsilon^{(n+1)} < \varepsilon^{(n)}$, that is $\{\varepsilon^{(n)}\}_{n=0}^{\infty} \subset [0, 1)$ is strictly decreasing, and

$$\left|\lambda_{\max}\left(B^{(n)^T}B^{(n)}\right)\right| < 1 - \varepsilon^{(n+1)}; \quad \forall n \in \mathbf{Z}_{0+} \tag{A11}$$

The constrains (A10), (A11) are jointly fulfilled if $\left|\lambda_{\max}\left(B^{(n)^T}B^{(n)}\right)\right| < \min\left(1 - \varepsilon^{(n+1)}, \varepsilon^{(n)} - \varepsilon^{(n+1)}\right)$; $\forall n \in \mathbf{Z}_{0+}$. □

Proof of Lemma 9. Since $A_m = EA = IEA$ then $vec(EA) = vec(IEA) = (I \otimes A^T) vec(E)$, and (32) becomes for I being the identity matrix of the same order as E and A:

$$(B \otimes C^T) vec(K) = vec((E-I)A) = (I \otimes A^T) vec(E) - vec(A) \tag{A12}$$

whose solutions are given by (34) if (33) holds and the whole set of solutions reduces to (35) if the solution is unique. The corresponding Equations (36)–(38) are got for the particular case when $A_m = \rho A$ with $|\rho| < 1/\|A\|_2$. Property (i) has been proved. Property (ii) follows since the least-squares best approximation to the corresponding incompatible algebraic system (31), or (32), is (40), that is, (34) for $k_w = 0$, [13,14]. □

Proof of Theorem 2. The solvability of (41) in the form (42) follows from the Rouché-Froebenius rank conditions from the algebraic compatibility under Assumptions 1,2. By defining

$$A^{0(n+1)} = \left[A^{(n)} \; 0_{(\sum_{i=0}^{n+1} n_i) \times n_{n+1}} \right]; \; B^{0(n+1)} = B^{(n)}$$

in order to complete a square "a priori" matrix of dynamics of the $(n+1)$-the aggregated system obtained after the aggregation of the $(n+1)$-th subsystem, note that

$$A^{0(n+1)} = \begin{bmatrix} A^{0(n)} + B_0^{(n)} K_0^{(n)} C^{(n)} & \hat{A}^{0(n)} & 0 \\ \hat{A}^{(n+1)} + \hat{B}_0^{(n)} \hat{K}_0^{a(n)} C^{(n)} & \hat{D}^{(n)} + \widetilde{\hat{D}}^{(n)} + \hat{B}_0^{(n)} \hat{K}_0^{(n)} \hat{C}^{(n)} & 0 \end{bmatrix} \in R^{(\sum_{i=0}^{n+1} n_i) \times (\sum_{i=0}^{n+1} n_i)} \tag{A13}$$

$$B^{0(n+1)} = \begin{bmatrix} \left(B_1^{(n)} + B_0^{(n)} K_1^{(n)}\right) C^{(n-1)} & 0 & \cdots & \left(B_j^{(n)} + B_0^{(n)} K_r^{(n)}\right) C^{(n-r)} & 0 \\ \left(\hat{B}_1^{a(n)} + \hat{B}_0^{(n)} \hat{K}_1^{a(n)}\right) C^{(n-1)} & \left(\hat{B}_1^{(n)} + \hat{B}_0^{(n)} \hat{K}_1^{(n)}\right) \hat{C}^{(n-1)} & \cdots & \left(\hat{B}_j^{a(n)} + \hat{B}_0^{(n)} \hat{K}_r^{a(n)}\right) C^{(n-r)} & \left(\hat{B}_j^{(n)} + \hat{B}_0^{(n)} \hat{K}_r^{(n)}\right) \hat{C}^{(n-r)} \end{bmatrix} \tag{A14}$$

$$\in R^{(\sum_{i=0}^{n+1} n_i) \times (\sum_{j=n-r}^{n-1} (\sum_{i=0}^{j+1} n_i))}$$

$\forall n \in Z_{0+}$. From (41), with corresponding associated controller explicit solutions (42), one gets that the $(n+1)$-th aggregated delay-free dynamics is described by the matrix:

$$A^{0(n+1)} = \left[A^{(n)} \; 0_{(\sum_{i=0}^{n+1} n_i) \times n_{n+1}} \right] = \begin{bmatrix} A_f^{(n)} & \hat{A}_f^{0(n)} & 0 \\ \hat{A}_f^{a(n+1)} & \hat{A}_f^{(n)} & 0 \end{bmatrix} \in R^{(\sum_{i=0}^{n+1} n_i) \times (\sum_{i=0}^{n+1} n_i)}; \; \forall n \in Z_{0+} \tag{A15}$$

Having in mind (27), construct

$$M^{(n+1)} = \overline{A}^{(n)} \overline{A}^{(n)T} = \begin{bmatrix} A_f^{(n)} & \hat{A}_f^{0(n)} & 0 & \overline{B}_1^{0(n)} \\ \hat{A}_f^{a(n+1)} & \hat{A}_f^{(n)} & 0 & \overline{B}_2^{0(n)} \\ i\left(\overline{B}^{0(n)}\right)I & & 0 & \end{bmatrix} \begin{bmatrix} A_f^{(n)} & \hat{A}_f^{0(n)} & 0 & \overline{B}_1^{0(n)} \\ \hat{A}_f^{a(n+1)} & \hat{A}_f^{(n)} & 0 & \overline{B}_2^{0(n)} \\ i\left(\overline{B}^{0(n)}\right)I & & 0 & \end{bmatrix}^T$$

$$= \begin{bmatrix} A_f^{(n)} A_f^{(n)T} + \hat{A}^{0(n)} \hat{A}^{0(n)T} + \overline{B}_1^{0(n)} \overline{B}_1^{0(n)T} & A_f^{(n)} \hat{A}_f^{a(n+1)T} + \hat{A}^{0(n)} \hat{A}_f^{(n)T} + \overline{B}_1^{0(n)} \overline{B}_2^{0(n)T} & \begin{bmatrix} A_f^{(n)} & \hat{A}_f^{0(n)} \\ \hat{A}_f^{a(n+1)} & \hat{A}_f^{(n)} \end{bmatrix} \\ \begin{bmatrix} A_f^{(n)} & \hat{A}_f^{0(n)} \\ \hat{A}_f^{a(n+1)} & \hat{A}_f^{(n)} \end{bmatrix}^T & & i\left(\overline{B}^{0(n)}\right)I \end{bmatrix}$$

$$= \begin{bmatrix} A_f^{(n)} A_f^{(n)T} + \hat{A}^{0(n)} \hat{A}^{0(n)T} + \overline{B}_1^{0(n)} \overline{B}_1^{0(n)T} & A_f^{(n)} \hat{A}_f^{a(n+1)T} + \hat{A}^{0(n)} \hat{A}_f^{(n)T} + \overline{B}_1^{0(n)} \overline{B}_2^{0(n)T} & \begin{bmatrix} A_f^{(n)} & \hat{A}_f^{0(n)} \\ \hat{A}_f^{a(n+1)} & \hat{A}_f^{(n)} \end{bmatrix} \\ \begin{bmatrix} A_f^{(n)T} & \hat{A}_f^{a(n+1)T} \\ \hat{A}_f^{0(n)T} & \hat{A}_f^{(n)T} \end{bmatrix} & & i\left(\overline{B}^{0(n)}\right) \delta^{(n)} I \end{bmatrix}$$

$$+ i\left(\overline{B}^{0(n)}\right) \begin{bmatrix} 0 & 0 \\ 0 & (1-\delta^{(n)})I \end{bmatrix}$$

where

$$\overline{B}^{0(n)} = \begin{bmatrix} \overline{B}_1^{0(n)} \\ \overline{B}_2^{0(n)} \end{bmatrix}$$

$$= \begin{bmatrix} \left(B_1^{(n)} + B_0^{(n)} K_1^{(n)}\right) \hat{C}^{(n-1)} & 0 & \cdots & \left(B_r^{(n)} + B_0^{(n)} K_r^{(n)}\right) \hat{C}^{(n-r)} & 0 \\ \left(\hat{B}_1^{a(n)} + \hat{B}_0^{(n)} \hat{K}_1^{a(n)}\right) \hat{C}^{(n-1)} & \left(\hat{B}_1^{(n)} + \hat{B}_0^{(n)} \hat{K}_1^{(n)}\right) \hat{C}^{(n-1)} & \cdots & \left(\hat{B}_r^{a(n)} + \hat{B}_0^{(n)} \hat{K}_r^{a(n)}\right) \hat{C}^{(n-r)} & \left(\hat{B}_r^{(n)} + \hat{B}_0^{(n)} \hat{K}_r^{(n)}\right) \hat{C}^{(n-r)} \end{bmatrix} \tag{A16}$$

and

(a) $i\left(\overline{B}^{0(n)}\right)$ is a binary indicator function defined as $i\left(\overline{B}^{0(n)}\right) = 1$ if $\left(\overline{B}^{0(n)}\right) \neq 0$ and $i\left(\overline{B}^{0(n)}\right) = 0$ if $\left(\overline{B}^{0(n)}\right) = 0$. The reason of the use of this indicator is that, in fact, if the delayed dynamics is zero then the dimension of the extended state, so that of $\overline{A}^{(n)}$, decreases since the resulting block identity matrices are removed,

(b) $\{\delta^{(n)}\}_{n=0}^{\infty} \subset [0, 1]$ is a design sequence which satisfies $\{\delta^{(n)}\}_{n=0}^{\infty} \to 0$.

Note that the fact that $I = i\left(\overline{B}^{0(n)}\right) I = i\left(\overline{B}^{0(n)}\right) \delta^{(n)} I + i\left(\overline{B}^{0(n)}\right) \left(1 - \delta^{(n)}\right) I$ justifies (A16). □

References

1. Ibeas, A.; De La Sen, M. Robustly stable adaptive control of a tandem of master–slave robotic manipulators with force reflection by using a multiestimation scheme. *IEEE Trans. Syst. Man Cybern. Part B (Cybernetics)* **2006**, *36*, 1162–1179. [CrossRef]
2. Barambones, O.; Garrido, A.J.; Garrido, I. Robust speed estimation and control of an induction motor drive based on artificial neural networks. *Int. J. Adapt. Control Signal Process.* **2008**, *22*, 440–464. [CrossRef]
3. Bakule, L.; de la Sen, M. Decentralized stabilization of networked complex composite systems with nonlinear perturbations. In Proceedings of the 2009 IEEE International Conference on Control and Automation, Christchurch, New Zealand, 9–11 December 2009; pp. 2272–2277.
4. Singh, M.G. *Decentralised Control*; North-Holland: Systems and Control Series; North Holland Publishing Company: New York, NY, USA, 1981; Volume 1.
5. Berman, A.; Plemmons, R.J. *Nonnegative Matrices in the Mathematical Sciences*; Academic Press: New York, NY, USA, 1979.
6. Kailath, T. *Linear Systems*; Prentice-Hall Inc.: Englewood Cliffs, NJ, USA, 1980.
7. Ortega, J.M. *Numerical Analysis*; Academic Press: New York, NY, USA, 1972.
8. Kouachi, S. The Cauchy interlace theorem for symmetrizable matrices. *arXiv Preprint* **2016**, arXiv:1603.04151.
9. Almutairi, S.; Thenmozhi, V.; Watkins, J.; Sawan, M.E. Optimal Control Design of Large Scale Systems with Uncertainty. In Proceedings of the 2018 IEEE 61st International Midwest Symposium on Circuits and Systems (MWSCAS), Windsor, ON, Canada, 5–8 August 2018; pp. 372–375.
10. Auger, P.; de La Parra, R.B.; Poggiale, J.C.; Sánchez, E.; Sanz, L. Aggregation methods in dynamical systems and applications in population and community dynamics. *Phys. Life Rev.* **2008**, *5*, 79–105. [CrossRef]
11. Aoki, M. Control of large-scale dynamic systems by aggregation. *IEEE Trans. Autom. Control* **1968**, *13*, 246–253. [CrossRef]
12. Darbha, S.; Rajagopal, K.R. Aggregation of a class of linear interconnected systems. In Proceedings of the American Control Conference, San Diego, CA, USA, 2–4 June 1999; pp. 1496–1501.
13. Mackenroth, U. *Robust Control Systems. Theory and Case Studies*; Springer: Berling/Heidelberg, Germany, 2003.
14. Barnett, S. *Matrices in Control Theory with Applications to Linear Programming*; Van Nostrand Reinhold Company: London, UK, 1971.
15. Verma, R.; Sehgal, V.K. Computational stochastic modelling to handle the crisis occurred during community epidemic. *Ann. Data Sci.* **2016**, *3*, 119–133. [CrossRef]
16. Iggidr, A.; Souza, M.O. State estimators for some epidemiological systems. *J. Math. Biol.* **2019**, *78*, 225–256. [CrossRef] [PubMed]
17. Yang, H.M.; Freitas, A.R. Biological view of vaccination described by mathematical modellings: from rubella to dengue vaccines. *Math. Biosci. Eng.* **2019**, *16*, 3195–3214. [CrossRef]

18. De la Sen, M.; Agarwal, R.P.; Ibeas, A.; Alonso-Quesada, S. On a generalized time-varying SEIR epidemic model with mixed point and distributed time-varying delays and combined regular and impulsive vaccination controls. *Adv. Differ. Equ.* **2010**, *1*, 1–42. [CrossRef]
19. De la Sen, M.; Ibeas, A.; Alonso-Quesada, S.; Nistal, R. On a SIR model in a patchy environment under constant and feedback decentralized controls with asymmetric parameterizations. *Symmetry* **2019**, *11*, 430. [CrossRef]
20. Hu, S.A.; Smith, R.L. The Schur complement interlacing theorem. *SIAM J. Matrix Anal. Appl.* **1995**, *16*, 1013–1023. [CrossRef]
21. Mercer, A.M.; Mercer, P.R. Cauchy's interlace theorem and lower bounds for the spectral radius. *Int. J. Math. Math. Sci.* **2000**, *23*, 563–566. [CrossRef]
22. Hwang, S.G. Cauchy's interlace theorem for eigenvalues of Hermitian matrices. *Am. Math. Mon.* **2004**, *111*, 157–159. [CrossRef]

© 2019 by the author. Licensee MDPI, Basel, Switzerland. This article is an open access article distributed under the terms and conditions of the Creative Commons Attribution (CC BY) license (http://creativecommons.org/licenses/by/4.0/).

Article

A Dynamic Simulation of the Immune System Response to Inhibit And Eliminate Abnormal Cells

S. A. Alharbi [1] and A. S. Rambely [2],*

[1] Department of Mathematics & Statistics, College of Science, Taibah University, 41911 Yanbu, Almadinah Almunawarah, Saudi Arabia; s.n.a.2013@hotmail.com
[2] Mathematics Program, Faculty of Science & Technology, Universiti Kebangsaan Malaysia, 43600 UKM Bangi, Selangor, Malaysia
* Correspondence: asr@ukm.edu.my

Received: 15 March 2019; Accepted: 12 April 2019; Published: 19 April 2019

Abstract: Diet has long been considered a risk factor related to an increased risk of cancer. This challenges us to understand the relationship between the immune system and diet when abnormal cells appear in a tissue. In this paper, we propose and analyze a model from the point of view of a person who follows a healthy diet, i.e., one correlated to the food pyramid, and a person who follows an unhealthy diet. Normal cells and immune cells are used in the design of the model, which aims to describe how the immune system functions when abnormal cells appear in a tissue. The results show that the immune system is able to inhibit and eliminate abnormal cells through the three following stages: the response stage, the interaction stage, and the recovery stage. Specifically, the failure of the immune system to accomplish the interaction stage occurs when a person follows an unhealthy diet. According to the analysis and simulation of our model, we can deduce that dietary pattern has a significant impact on the functioning of the immune system.

Keywords: dynamic model; immune system response; immune cells; abnormal cells; nonlinear ordinary differential equations; stability; diet

1. Introduction

Lifestyle has changed significantly in recent decades due to rapid development in every sphere of life. Consequently, the rate of noncommunicable diseases (NCDs), such as cardiovascular disease, cancer, chronic respiratory disease, and diabetes, has dramatically increased. In 2016, the World Health Organization (WHO) reported that 71% of all deaths worldwide occurred as a result of NCDs. According to the report, the highest mortality figures were those related to cardiovascular disease, representing 44% of deaths from the four main NCDs. The second most deadly disease was cancer, accounting for approximately 22% of deaths from the four main NCDs; while chronic respiratory disease and diabetes reached around 9% and 4%, respectively. In addition, gender variation in NCDs is another important factor. Research found that adult men are more likely to be affected than adult women, with 22% of men and 15% of women being affected [1]. According to a recent study, cancer is more common in certain countries; for example, Australia reported the highest percentage of cancer, with 4680 people per 100,000 having the disease. New Zealand registered 4381 people per 100,000. The number of cancer cases has been estimated as 3522 per 100,000 in the USA, compared with 3192 registered cases per 100,000 in the United Kingdom [2]. The tissues and organs of the human body are formed from 10^{13} tiny cells. There is a one-to-one correspondence between cells and human body growth. The more increased the number of cells, the more tissue grows. The cells between conception and adulthood divide and grow very quickly [3]. Yet, the functions of these cells vary, and as a result the division and growth of the cells depend on their functions. For example, blood and skin cells divide continuously, while some cells have particular functions in the body and do not usually multiply.

Concerning the multiplication of cells, it is possible for them to multiply as many as 60 times before dying, as a result of the signals that control cellular growth and death [4,5]. On the other hand, they can become damaged during the process of division, which can lead to self-elimination. This process is known as apoptosis and it protects the human body from cancer. Conversely, cell division is sometimes abnormal when there is damage during cell division, with very unique characteristics [3–5]. In such cases, the immune system reacts to protect the human body by preventing these cells from growing into a tumor [3–5]. Assessing modern lifestyles is the key to understanding the causes of the increasing rates of cancer. One of the disadvantages of the development of society is that dietary habits have changed to encompass more fast and processed foods. These types of foods have a higher calorie count, contain more protein, and have lower amounts of fiber and carbohydrates than healthy foods, which are rich in natural sources of vitamins and minerals and high in fiber and carbohydrates [6]. Studies have shown the increase in death rate associated with a diet based on animal products and a high intake of carbohydrates, and contrarily, how a vegetable-based diet and a low carbohydrate intake reduces the mortality rate. Furthermore, malnutrition debilitates the immune system and increases the mortality rate as well as elevating the risk of contracting NDCs [7]. It has been shown that only 5–10% of cancers occur as a result of internal factors such as inherited mutations, hormones, and immune conditions, and that 90–95% of cancers are due to lifestyle and environmental factors [8]. Dietary habits are one of the main factors related to the weakening of the immune system and the risk of cancer [9–12]. From 1994, mathematical researchers started to formulate tumor–immune interaction models using the function of Michaelis–Menten [13,14]. In 1995, Mayer and others proposed a basic mathematical model of the immune response by using two ordinary differential equations to describe the interaction between the immune system and a pathogen, such as a tumor cell or virus. Their model succeeded in illustrating that the combination of a few proposed nonlinear interaction rules between the immune system and pathogens is able to generate a considerable variety of immune responses, with many of them being observed both experimentally and clinically. Hence, the process of the interaction of the immune system with pathogens can be described dynamically [15]. In 2003, Magda Galach used the simplified model of Kuznetsov–Taylor and changed the Michaelis–Menten function using the Lotka–Volterra form [16]. Many models have used ordinary differential equations, partial differential equations, and delay differential equations to illustrate the growth of tumors and their treatments [17–19]. In the last few years, mathematical researchers have dynamically examined cancer risk factors and yet still have a great deal to uncover. Estrogen has been studied as a breast cancer risk [20]; furthermore, Green and others studied the relationship between body mass index, menopausal status, estrogen replacement therapy and the risk of breast cancer [21]. In 2016, Roberto and others proposed an obesity–cancer model using ordinary differential equations to illustrate the association between obesity and cancer risk [22]. Additionally, they presented the effects of obesity on the optimal control program of chemotherapy. This model differs from that of De Pillis and Radunskaya [23] by adding a dynamic equation related to stored fat [24]. Other studies have presented models regarding drug therapy which contribute to decision making and early cancer treatment [25]. In 2018, Alharbi and Rambely suggested dynamically that switching back to a healthy lifestyle boosted the immune system in terms of inhibiting or eliminating a moderately abnormal cell [26]. Hence, the immune system has the ability to protect the human body from developing cancer.

In 1618, Thomas Adams stated that "Prevention is so much better than healing because it saves the labour of being sick" [27]. In this work, we propose and simulate the immune–healthy diet model (IHDM) based on the models of references [15,26] using ordinary differential equations to study the behavior of the immune system when responding to the appearance of abnormal cells which fail to eliminate themselves. Usually, these types of cells do not need to be treated clinically. However, the appearance of abnormal cells in the tissue is considered as an emergency situation, with any progression being able to trigger the formation of tumor cells and the development of cancer. Therefore, most cancers develop as a consequence of multiple abnormalities, which accumulate over many years [5]. For instance, colon cancer has increased more than tenfold between the ages of 30 and 50, and tenfold again between 50 and 70 [28].

We aim to enhance understanding of the symmetry and antisymmetry of the relationship between diet habits and the function of the immune system, raise awareness of healthy habits and promote healthy eating habits. In doing so, we hope that the results of this paper contribute to increasing awareness of cancer risk.

This paper is organized as follows. In Section 2, we present the IHDM and analyze the equilibrium points of the model and stability cases. In Section 3, we analyze the stability of the equilibrium points for the immune–unhealthy diet model (IUNHDM). The numerical simulation of the IHDM and IUNHDM are presented in Section 4. The conclusion is presented in Section 5.

2. The Immune—Healthy Diet Model (IHDM)

The immune system is very complex. One of its main functions is to recognize pathogens and to protect the body from developing diseases. However, the response of the immune system is affected by dietary habits, physical activity, stress, and sleep habits. A balanced diet and adequate intake of vitamins act to boost the immune system [29].

Our model is composed of ordinary differential equations, and it supposes that the individual follows a healthy diet as per the food pyramid, which is shown in Figure 1. This food pyramid follows recommendations from a report from the World Cancer Research Fund (WCRF) and the American Institute for Cancer Research (AICR), which recommends drinking water, eating a diet rich in wholegrains, vegetables, fruits, and beans, and having a lower intake of red and processed meats, as well as sugars and sweets [30,31]. The IHDM is formulated into two main populations: normal cells and immune cells activated as a result of the inability of the abnormal cells to eliminate themselves automatically. According to a cell's life cycle, we formulated the first equation to describe the behavior of normal cells during the process of their dynamic division and growth and to consider that some cells might divide abnormally. In addition, the natural death of the normal cells where the elimination of these cells occurs by apoptosis was not considered. The dynamic process of the interaction between the immune cells with the abnormal cells is presented in the second equation. This considers abnormal cells in the primary stage, which is where most cancers develop as a consequence of multiple abnormalities accumulating over many years. At this stage, the immune system eliminates them from growing before they turn into tumor cells (by attacking and repairing processes).

The IHDM is given as follows:

$$\begin{aligned} \frac{dN}{dt} &= rN[1-\beta N] - \eta NI, \\ \frac{dI}{dt} &= \sigma - \delta I - \frac{\rho NI}{m+N} - \mu NI, \end{aligned} \quad (1)$$

with initial values $N(0) = 1$ and $I(0) = 1.22$, where the dependent variables N and I represent the population of normal cells and immune cells, respectively. The parameters $r, \beta, \eta, \sigma, \delta, \rho, m$, and μ are real and positive. The rate of growth of normal cells is represented by the parameter r and the rate of appearance of abnormal cells during the cell life cycle of the normal cell is given by β. Furthermore, the fixed source of immune cells is represented by σ and their rate of natural death is represented by δ. The ability of immune cells to eliminate abnormal cells determined by the Michaelis–Menten term $\frac{\rho IN}{m+N}$ [15]. The coexistence of abnormal cells stimulates the immune system to respond [20]. The rate of this response is represented by ρ and the parameter m represents the threshold rate of the immune system. The parameters η and μ display the interaction between the abnormal cells and immune cells. The ability of the immune cells to eliminate abnormal cells or inhibit them is given by parameter η, whereas the parameter μ illustrates the decreasing number of immune cells as a result of their interaction with abnormal cells. In the IHDM, the rate of the parameter $\eta > \mu$ represents the case where the immune system is strong and succeeding in performing its function and the person is following a healthy diet [29].

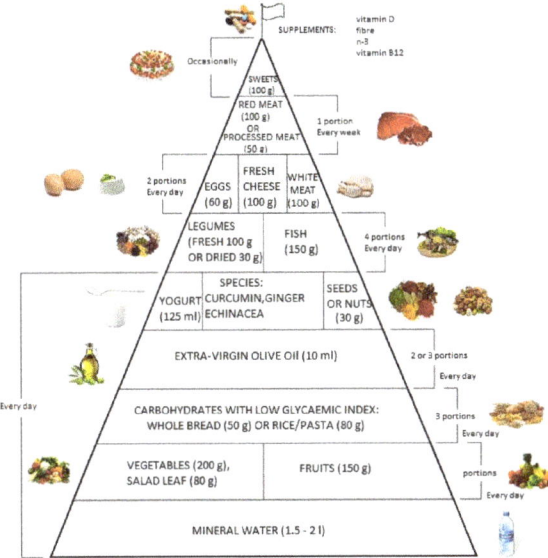

Figure 1. The dietary management food pyramid according to the World Cancer Research Fund (WCRF) and American Institute for Cancer Research (AICR) where the amounts of food are estimated based on nutritional and practical considerations.

2.1. Equilibrium Points

In this section, we use nullclines from system (1) to compute the equilibrium points as follows:

1. $\frac{dN}{dt} = 0 \Leftrightarrow$

$$N = 0,$$
$$r[1 - \beta N] - \eta I = 0.$$

2. $\frac{dI}{dt} = 0 \Leftrightarrow$

$$\sigma - \delta I - \frac{\rho N I}{m + N} - \mu N I = 0.$$

We classify the equilibrium points according to their biological terms as the following:

1. Primary response stage: The immune system recognizes the appearance of abnormal cells in the tissue, immune cells start to grow, and $N = 0$; this point is given by $p_1 = (0, \frac{\sigma}{\delta})$.
2. Interaction stage: The immune cells eliminate or inhibit the abnormal cells, this means $\beta \to 0$ at the end of the interaction. This point is given by

$$p_2 = (\frac{-r(\delta + m\mu - \rho) + \eta\sigma + \sqrt{\Delta}}{2r\mu}, \frac{r}{\eta}),$$

where $\Delta = r(\delta + m\mu - \rho)^2 + 4mr\mu\delta(-r + \eta\frac{\sigma}{\delta}) > 0$.
3. Recovery stage: Immune cells that are involved in the reaction tend to zero and all abnormal cells are substituted with normal cells. This point is represented by

$$p_3 = (\beta^{-1}, \frac{\beta(1+m\beta)\sigma}{(1+m\beta)(\beta\delta+\mu)-\beta\rho}), \text{ where } 0 < \beta < 0.1.$$

2.2. Analysis Stability of Equilibrium Points

According to the concept of the Hartman–Grobman theorem, the hyperbolic equilibrium point in the neighborhood of a nonlinear dynamical system is topologically equivalent to its linearization. Therefore, the Jacobian of the nonlinear dynamic of (1) is computed as follows:

$$J[N, I] = \begin{bmatrix} F_N[N, I] & F_I[N, I] \\ G_N[N, I] & G_I[N, I] \end{bmatrix}, \quad (2)$$

where $F[N, I] = \frac{dN}{dt}$ and $G[N, I] = \frac{dI}{dt}$. The stability cases according to the above equilibrium points are as follows:

1. Stability of primary response stage- p_1: Under the hypothesis of the IHDM, the immune system is able to protect the human body from developing diseases. This means that it responds directly in cases of emergency such as the appearance of abnormal cells in the tissue. The Jacobian (2) at equilibrium point p_1 is computed as:

$$J[N, I]_{p_1} = \begin{pmatrix} r - \eta \frac{\sigma}{\delta} & -0 \\ \frac{m\rho\sigma}{m^2\delta} - \mu\frac{\sigma}{\delta} & -\delta \end{pmatrix}. \quad (3)$$

Proposition 1. *Since the immune system is strong, the equilibrium point p_1 is a stable node.*

Proof. From the Jacobian (3), the eigenvalues are given by

$$\lambda_1 = -\delta < 0,$$

$$\lambda_2 = r - \eta\frac{\sigma}{\delta} = r - \eta I(0) < 0 \Leftrightarrow I(0) > \frac{r}{\eta},$$

where $0 < \delta < 1$, then,

$$\delta^{-1} \geq 1 \Rightarrow \lambda_2 < \lambda_1 < 0.$$

Hence, the equilibrium point p_1 is a stable node, see Figure 2. □

2. Stability of interaction stage- p_2: This stage describes the ability of the immune cells to inhibit and eliminate the abnormal cells to prevent them from progressing to cancer over many years. We consider the model as having a significant interaction if abnormal cells are dying or being inhibited by the immune cells. This means that parameter $\beta \to 0$ at $t \to \infty$. To examine the stability of equilibrium point $p_2 = (\frac{-r(\delta + m\mu - \rho) + \eta\sigma\sqrt{\Delta}}{2r\mu}, \frac{r}{\eta})$, we compute the Jacobian (2) at this point as

$$J[N, I]_{p_2} = \begin{pmatrix} \frac{-\beta}{\mu}x & \frac{-\eta}{2r\mu}x \\ \frac{r\mu}{\eta}(-1 + \frac{4m\mu\rho r^2}{y^2}) & \frac{-\eta\sigma}{r} \end{pmatrix}, \quad (4)$$

where

$$x = -r(\delta + m\mu - \rho) + \eta\sigma + \sqrt{\Delta},$$
$$y = -r\delta + rm\mu + r\rho + \eta\sigma + \sqrt{\Delta}.$$

Proposition 2. *There is an unstable saddle equilibrium point during the interaction stage.*

Proof. From the Jacobian (4), the characterized equation is given by

$$(\frac{-\beta}{\mu}x - \lambda)(\frac{-\eta\sigma}{r} - \lambda) - \frac{r\mu}{\eta}(-1 + \frac{4m\mu\rho r^2}{y^2})(\frac{\eta}{2r\mu}x) = 0 \quad (5)$$

$$(\frac{-\beta}{\mu}x - \lambda)(\frac{-\eta\sigma}{r} - \lambda) - (-1 + \frac{4m\mu\rho r^2}{y^2})(\frac{1}{2}x) = 0. \qquad (6)$$

We assume that $C_1 = \frac{x}{2}(-1 + \frac{4m\mu\rho r^2}{y^2}) > 0$. Then, we rewrite (6) as follows:

$$\lambda^2 + (\frac{\beta x}{\mu} + \frac{\eta\sigma}{r})\lambda + (\frac{-\eta\sigma}{r})(\frac{\beta x}{\mu}) - C_1 = 0. \qquad (7)$$

Since, under the hypotheses of the IHDM, $\beta \to 0$ at $t \to \infty$, Equation (7) is computed as

$$\lambda^2 + \frac{\eta\sigma}{r}\lambda - C_1 = 0. \qquad (8)$$

Now, we apply the Routh–Hurwitz theorem for (8), giving

$$\begin{vmatrix} \lambda^2 & 1 & -C_1 \\ \lambda^1 & \frac{\eta\sigma}{r} & 0 \\ \lambda^0 & -C_1 & 0 \end{vmatrix}.$$

Since the first column has one sign change, the equilibrium point p_2 is an unstable saddle. □

3. **Stability of recovery stage-** p_3: According to the physiological process, the number of immune cells which are involved in the interaction starts to reduce automatically after inhibiting and eliminating the abnormal cells. Furthermore, the normal cells divide and grow, taking the place of the removed abnormal cells. To examine the stability of this point, we compute the Jacobian at $p_3 = (\beta^{-1}, \frac{\beta(1+m\beta)\sigma}{(1+m\beta)(\beta\delta+\mu)-\beta\rho})$ as follows:

$$J[N, I]_{p_3} = \begin{pmatrix} \frac{-rz + \beta\eta(1+m\beta)\sigma}{z} & \frac{-\eta}{\beta} \\ \frac{\beta(-(1+m\beta)^2\mu + m\beta^2\rho)\sigma}{z(1+m\beta)} & -\frac{z}{\beta+m\beta^2} \end{pmatrix}, \qquad (9)$$

where $z = \beta(\delta + m\mu - \rho) + m\delta\beta^2 + \beta\eta(1 + m\beta)\sigma > 0$.

Proposition 3. *In the recovery stage, the system might to be stable.*

Proof. The characterized equation of (9) is given by

$$(\frac{-rz + \beta\eta(1+m\beta)\sigma}{z} - \lambda)(-\frac{z}{\beta+m\beta^2} - \lambda) - C_2 = 0, \qquad (10)$$

where

$$C_2 = (\frac{-\eta}{\beta})(\frac{\beta(-(1+m\beta)^2\mu + m\beta^2\rho)\sigma}{z(1+m\beta)}) > 0.$$

To simplify, we let

$$A = \frac{-rz + \beta\eta(1+m\beta)\sigma}{z} - \frac{z}{\beta+m\beta^2} < 0,$$

and

$$B = (\frac{-rz + \beta\eta(1+m\beta)\sigma}{z})(-\frac{z}{\beta+m\beta^2}) > 0.$$

Then, Equation (10) is rewritten as

$$\lambda^2 - A\lambda + D = 0, \qquad (11)$$

where $D = B - C_2 > 0$. We apply the Routh–Hurwitz theorem for (11) to determine the sign of roots:

$$\begin{vmatrix} \lambda^2 & 1 & D \\ \lambda^1 & -A & 0 \\ \lambda^0 & D & 0 \end{vmatrix}.$$

Since $A < 0$, the sign of elements in first column is positive and the equilibrium point p_3 is called a stable node point for the IHDM. □

With reference to propositions (1,2,3), we conclude this section with the following remark:

Remark 1. *The IHDM has the following properties:*

- *The system has three equilibrium points;*
- *The system has two equilibrium points which are stable nodes, which shows that the immune system plays a pivotal role in protecting the human body from diseases.*
- *The system has only one equilibrium point which is an unstable saddle, which shows the interaction between the immune system and abnormal cells.*
- *The phase portrait of the IHDM and its solutions around the equilibrium points are shown in Figures 2 and 3.*

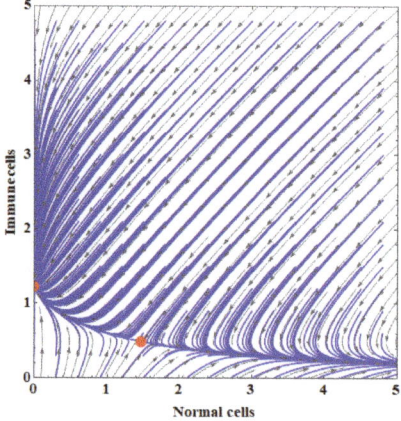

Figure 2. The phase portrait of the immune–healthy diet model (IHDM) and its solutions around the response and interaction equilibrium points.

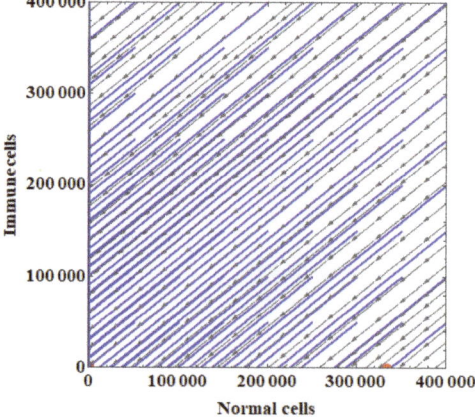

Figure 3. The phase portrait of the IHDM and its solutions around the recovery equilibrium points.

3. Immune-Unhealthy Diet Model (IUNHDM)

A Global Nutrition report (2016) stated that every third person in the world is inflicted with malnutrition [32]. On this basis, a pre-clinical experiment proved that high protein and fructose diets damage the immune system and cause cancer [33–35]. In this section, we highlight why a weak immune system fails to protect the human body from developing cancer; we do this using a similar system to that presented in (1) and under the assumption that the person follows an unhealthy diet. For instance, the western-style diet which is characterized by a high consumption of proteins (especially high in processed and red meats), sugar, salt, and fat when compared to a healthy diet comprising an intake of fruits and vegetables [36–38]. Several studies showed that a high sugar and protein diet damages the immune system and promotes cancer [33,34,39]. Hence, In the IUNHD, the effects of an unhealthy diet are presented dynamically according to the rate of two parameters: μ and η. In this section, we analyze the model considering an example where the immune system is weak as a result of the person following an unhealthy diet.

3.1. Equilibrium Points:

The model has only two equilibrium points in the feasible region, when it is compared to the IHDM. These equilibrium points are represented as follows:

1. Primary response stage: In this stage, the immune system recognizes abnormal cells as foreign; this point is represented by $u_1 = (0, \frac{\sigma}{\delta})$. This is a similar equilibrium point to that of the IHDM.
2. Coexistence stage: In this stage, the immune cells treat abnormal cells as normal cells; this point is represented by

$$u_2 = (\beta^{-1}, \frac{\beta(1+m\beta)\sigma}{(1+m\beta)(\beta\delta+\mu)-\beta\rho}), \text{ where } 0 < \beta < 0.1.$$

Remark 2. *Mathematically, both the recovery and the coexistence stages are the same, but their meaning in terms of physiology is totally different. The stage of recovery follows the interaction stage, which means that point p_3 represents the total of the population after the substitution of abnormal cells with normal cells. However, the point of u_2 in the IUNHDM may mathematically indicate the coexisting population of both normal cells and abnormal cells, as suggested by [20]; this type of coexistence might trigger cancer [40].*

Remark 3. *This model does not have an equilibrium point for the interaction stage; this means that one of the following is true:*

- *There was a response from the immune system but the immune cells did not become involved in the interaction because the immune system was weak, or;*
- *The immune cells became involved in the interaction but failed to inhibit or eliminate the abnormal cells. Hence, this type of interaction damages the immune cells. In other words, the population of $I \to$ zero before inhibiting or eliminating the abnormal cells.*

3.2. Stability of Equilibrium Points

1. Stability of primary response stage: Since this point is identical to the primary response stage for the IHDM, we use (3) to examine the stability of this stage for the IUNHDM.

Proposition 4. *The point which represents the primary response stage is the unstable node for the IUNHDM.*

Proof. On the basis of the weakness of the immune system, we can hypothesize that $\eta I < r$, where $I = \frac{\sigma}{\delta}$. That means that the parameter of η failed to achieve its highest peak rate. Therefore, the eigenvalues of matrix (3) have a different sign. Accordingly, the equilibrium point u_1 is an unstable node. □

2. Stability of coexistence stage: This stage is considered to be a trigger for cancer because the immune cells die without inhibiting or eliminating the abnormal cells. This era of rapid development affecting our lifestyle, especially our dietary habits, is one of the main causes of abnormal cells progressing early into tumor cells. Since this is mathematically similar to the recovery point in the IHDM, the stability case is given by the following proposition:

Proposition 5. *The point which represents the coexistence stage is stable.*

Remark 4. *The phase portrait of the IUNHDM and its solutions around the equilibrium points is shown in Figures 4 and 5.*

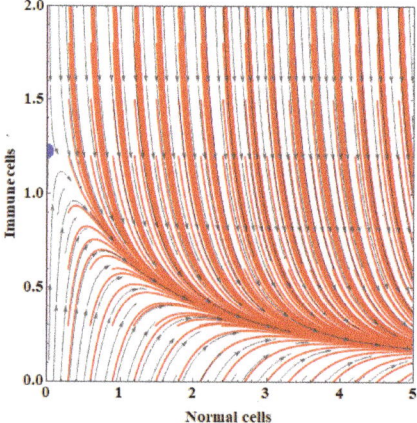

Figure 4. The phase portrait of the immune-unhealthy diet model (IUNHDM) and its solutions around the response equilibrium point.

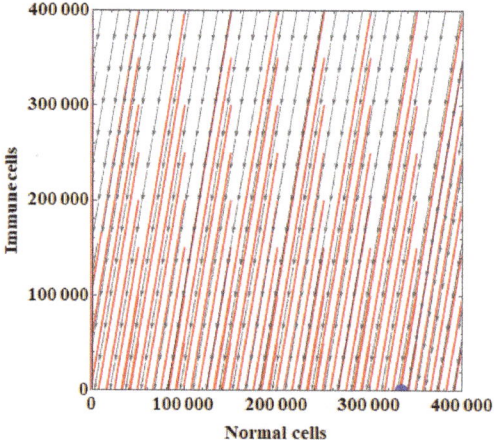

Figure 5. The phase portrait of the IUNHDM and its solutions around the coexistence equilibrium point.

4. Numerical Simulation

Both the IHDM and the IUNHDM were simulated using Mathematica software with the built-in functionality NDSolve. The proposed system of ordinary differential equations can be solved with any standard numerical method. For this, we used an explicit Runge–Kutta with a difference order of four and a step size of 1/10,000. The reliability and accuracy of the proposed numerical method can be seen from the residual error which is shown in Figures 6–9.

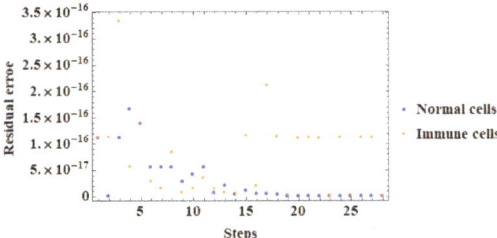

Figure 6. The residual error at steps for the proposed numerical method for the IHDM.

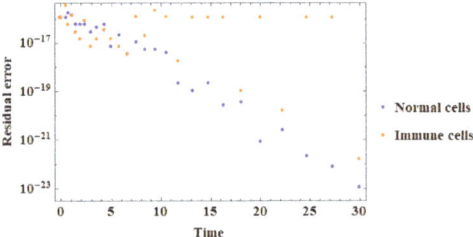

Figure 7. The residual error at time t for the proposed numerical method for the IHDM.

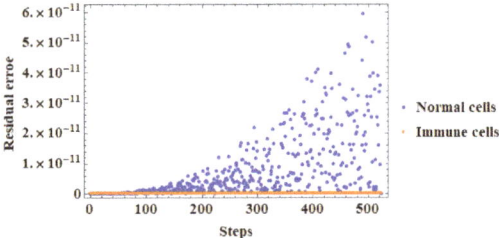

Figure 8. The residual error at steps for the proposed numerical method for the IUNHDM.

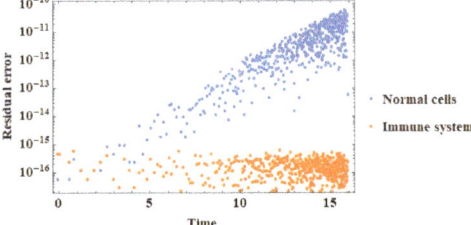

Figure 9. The residual error at time t for the proposed numerical method for the IUNHDM.

The simulation of the IHDM and the IUNHDM indicated that the immune system response is affected by specific parameters, namely, $\rho, m, \eta,$ and μ. These parameters are put as $m = 0.4787$, $\rho = 0.2206, \eta = 0.8791,$ and $\mu = 0.6986$ for the IHDM and $m = 0.3389, \rho = 0.2710, \eta = 0.1379$, and $\mu = 0.8130$ for the IUNHMD. In addition, the immune system can respond to an emergency case if and only if the threshold rate achieves its peak, which is given by 0.478; it is reduced to 0.3389 in the IUNHDM. This reveals a weakness in the immune system. Furthermore, the impact of the interaction between the immune system and abnormal cells depends on the rate of two parameters:

η and μ. By comparing both models, we can deduce that the immune system succeeded in performing its function if and only if

$$\text{the rate of parameter } \eta > \text{ the rate of parameter } \mu.$$

The behavior of the cells in the IHDM and the IUNHDM are shown in Figures 10 and 11, respectively. Furthermore, The numerical simulation revealed that in the IHDM, the immune system recognized abnormal cells as foreign bodies and started to inhibit or eliminate them within the first five days. In addition, the immune system put a copy of the abnormal cells in its memory, which helps it to eliminate them if they appear again or begin to progress. This is clearly seen as the population of immune cells returns to its initial value after the 10th day of interaction. Furthermore, the horizontal growth of normal cells indicates that normal cells divide and grow by following the signals of control cellular growth and death [5]. On the other hand, the response of the immune system and attitude of their cells were delayed in the IUNHDM. Hence, the immune cells failed to become involved in the interaction and reduced to zero, while the normal cells had a mutation which forced them to grow vertically, something which can lead to carcinoma [5,40].

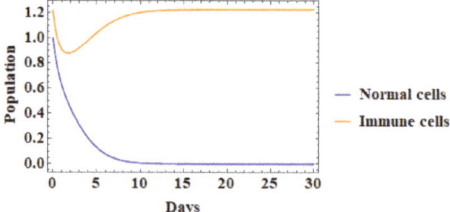

Figure 10. The behavior of the IHDM where $r = 0.431201, \beta = 2.99 \times 10^{-6}, \sigma = 0.7, \delta = 0.57, m = 0.4787, \rho = 0.2206, \eta = 0.8791,$ and $\mu = 0.6986$.

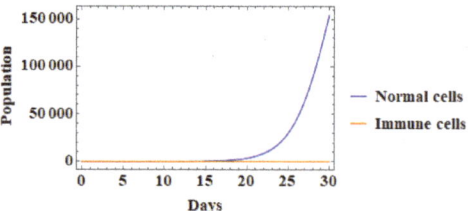

Figure 11. The behavior of the IUNHDM where $r = 0.431201, \beta = 2.99 \times 10^{-6}, \sigma = 0.7, \delta = 0.57, m = 0.3389, \rho = 0.2710, \eta = 0.1379,$ and $\mu = 0.8130$.

5. Conclusions

Alharbi and Rambely [26] formulated their model to examine the impact a modern lifestyle has on our health, as well as the impact of switching back from an unhealthy lifestyle to a healthy lifestyle. In this study, our model aims to understand the natural function of the immune system as regards protecting the human body from developing cancer where the progress of abnormal cells might trigger the appearance of tumor cells. Hence, the immune–healthy diet and immune–unhealthy diet models have been studied dynamically, analytically, and numerically. By comparing the results of the analysis and simulation of both the IHDM and IUNHDM, we suggested that there are three stages which act to stop abnormal cells from progressing into tumor cells. In the first stage, the immune system receives a signal which provokes it to recognize abnormal cells as foreign. Next, immune cells are activated to attack the abnormal cells. The results of this interaction lead to the inhibition or elimination of the abnormal cells. Finally, the immune cells typically die after the interaction stage, which indicates that the body is in the recovery stage. These processes were interrupted in the immune–unhealthy diet model, in which the interaction between the immune cells and abnormal cells failed to eliminate the abnormal cells and also led to a decrease in immune cells. Thereafter, the abnormal cells succeeded in coexisting with the normal cells; this type of coexistence might trigger cancer [5,20,22,24,26,40]. On the basis of the analysis and simulation of the IHDM and the IUNHDM, we can infer that the process of elimination or inhibition of abnormal cells by the immune system is affected by two main parameters, namely, η and μ. When a person follows a healthy diet (correlating with the food pyramid), his/her immune system will be strong and able to recognize damaged cells, responding by repairing or eliminating them. By contrast, following an unhealthy diet leads to a weakened immune system, which harms its function. As a result, abnormal cells will stimulate a response from the immune system, encouraging it to increase the generation of immune cells. This stimulation increased when associated with the IUNHDM and decreased when associated with IHDM. In summary, symmetry and antisymmetry are basic characteristics in the understanding of the relationship between dietary patterns and the behavioral responses of the immune system when protecting the human body from developing diseases. The symmetry of the IHDM and IUNHDM can be seen when the immune system in both models responds to abnormal cells appearing in the tissue, as well as the responses in the recovery stage. The results also suggested some similarities in terms of the function of the immune system in both models, possibly affected by diet habits. Although the mathematical model that is proposed in this work contributes to understanding the general dynamics of pathogens, it is well-known that mathematical models cannot take all variables into account. For this reason, it is highly recommended to conduct clinical experiments to consider real cases in order to confirm the results of our mathematical model and to show more precise results. In the future, we will expand this work to study the dynamic effect of the growth of abnormal cells and their activity. In addition, we plan to apply our work to other pathogens.

Author Contributions: Conceptualization, S.A.A.; Funding Acquisition, A.S.R.; Methodology, S.A.A.; Project Administration, A.S.R.; Supervision, A.S.R.; Validation, A.S.R.; Writing—Original Draft, S.A.A.; Writing—Review and Editing, A.S.R.

Funding: This research is funded by a grant from Universiti Kebangsaan Malaysia.

Acknowledgments: We are indebted to Universiti Kebangsaan Malaysia for providing financial support and facilities for this research under the grant GUP-2017-112.

Conflicts of Interest: The authors declare no conflict of interest.

References

1. World Health Organization. *World Health Statistics 2018: Monitoring Health for the SDGs Sustainable Development Goals*; World Health Organization: Geneva, Switzerland, 2018.
2. Bray, F.; Ferlay, J.; Soerjomataram, I.; Siegel, R.L.; Torre, L.A.; Jemal, A. Global cancer statistics 2018: GLOBOCAN estimates of incidence and mortality worldwide for 36 cancers in 185 countries. *Cancer J. Clin.* **2018**, *68*, 394–424. [CrossRef]
3. Howard, P.; Whittaker, B. *Placement Learning in Cancer & Palliative Care Nursing-E-Book: A Guide for Students in Practice*; Elsevier Health Sciences: Philadelphia, PA, USA, 2012.
4. Alberta, B.; Lewis, J.; Roberta, K.; Johnson, A.; Raff, M.; Walter, P. *Molecular Biology of the Cell*; Garland PUB: New York, NY, USA, 2008.
5. Cooper, G.M.; Hausman, R.E. *The Cell: A Molecular Approach*; ASM Press: Washington, DC, USA, 2000; Volume 2.
6. Boulet, L.P. Asthma and obesity. *Clin. Exp. Allergy* **2013**, *43*, 8–21. [CrossRef]
7. Fung, T.T.; van Dam, R.M.; Hankinson, S.E.; Stampfer, M.; Willett, W.C.; Hu, F.B. Low-carbohydrate diets and all-cause and cause-specific mortality: Two cohort studies. *Ann. Intern. Med.* **2010**, *153*, 289–298. [CrossRef]
8. Loeb, K.R.; Loeb, L.A. Significance of multiple mutations in cancer. *Carcinogenesis* **2000**, *21*, 379–385. [CrossRef] [PubMed]
9. Pal, D.; Banerjee, S.; Ghosh, A.K. Dietary-induced cancer prevention: An expanding research arena of emerging diet related to healthcare system. *J. Adv. Pharm. Technol. Res.* **2012**, *3*, 16.

10. Yusof, A.S.; Isa, Z.M.; Shah, S.A. Dietary patterns and risk of colorectal cancer: A systematic review of cohort studies (2000–2011). *Asian Pac. J. Cancer Prev.* **2012**, *13*, 4713–4717. [CrossRef] [PubMed]
11. Yusof, A.S.; Isa, Z.M.; Shah, S.A. Perceptions of Malaysian colorectal cancer patients regarding dietary intake: A qualitative exploration. *Asian Pac. J. Cancer Prev.* **2013**, *14*, 1151–1154. [CrossRef]
12. Shahar, S.; Shafurah, S.; Hasan Shaari, N.S.; Rajikan, R.; Rajab, N.F.; Golkhalkhali, B.; Zainuddin, Z.M.D. Roles of diet, lifetime physical activity and oxidative DNA damage in the occurrence of prostate cancer among men in Klang Valley, Malaysia. *Asian Pac. J. Cancer Prev.* **2011**, *12*, 605–611.
13. Kuznetsov, V.A.; Makalkin, I.A.; Taylor, M.A.; Perelson, A.S. Nonlinear dynamics of immunogenic tumors: Parameter estimation and global bifurcation analysis. *Bull. Math. Biol.* **1994**, *56*, 295–321. [CrossRef] [PubMed]
14. Kirschner, D.; Panetta, J.C. Modeling immunotherapy of the tumor–immune interaction. *J. Math. Biol.* **1998**, *37*, 235–252. [CrossRef]
15. Mayer, H.; Zaenker, K.; An Der Heiden, U. A basic mathematical model of the immune response. *Chaos Interdiscip. J. Nonlinear Sci.* **1995**, *5*, 155–161. [CrossRef]
16. Gałach, M. Dynamics of the Tumor—Immune System Competition—The Effect of Time Delay. *Int. J. Appl. Math. Comput. Sci.* **2003**, *13*, 395–406.
17. Villasana, M.; Radunskaya, A. A delay differential equation model for tumor growth. *J. Math. Biol.* **2003**, *47*, 270–294. [CrossRef] [PubMed]
18. Rihan, F.; Rihan, N. Dynamics of Cancer-Immune System with External Treatment and Optimal Control. *J. Cancer Sci. Ther.* **2016**, *8*, 257–261. [CrossRef]
19. Bratus, A.; Samokhin, I.; Yegorov, I.; Yurchenko, D. Maximization of viability time in a mathematical model of cancer therapy. *Math. Biosci.* **2017**, *294*, 110–119. [CrossRef] [PubMed]
20. Mufudza, C.; Sorofa, W.; Chiyaka, E.T. Assessing the effects of estrogen on the dynamics of breast cancer. *Comput. Math. Methods Med.* **2012**, *2012*, 473572. [CrossRef]
21. Green, L.E.; Dinh, T.A.; Smith, R.A. An estrogen model: The relationship between body mass index, menopausal status, estrogen replacement therapy, and breast cancer risk. *Comput. Math. Methods Med.* **2012**, *2012*, 792375. [CrossRef] [PubMed]
22. Ku-Carrillo, R.A.; Delgadillo, S.E.; Chen-Charpentier, B. A mathematical model for the effect of obesity on cancer growth and on the immune system response. *Appl. Math. Model.* **2016**, *40*, 4908–4920. [CrossRef]
23. De Pillis, L.G.; Radunskaya, A. A mathematical tumor model with immune resistance and drug therapy: An optimal control approach. *Comput. Math. Methods Med.* **2001**, *3*, 79–100. [CrossRef]
24. Ku-Carrillo, R.A.; Delgadillo-Aleman, S.E.; Chen-Charpentier, B.M. Effects of the obesity on optimal control schedules of chemotherapy on a cancerous tumor. *J. Comput. Appl. Math.* **2017**, *309*, 603–610. [CrossRef]
25. Wu, J.; Tan, Y.; Chen, Z.; Zhao, M. Data Decision and Drug Therapy Based on Non-Small Cell Lung Cancer in a Big Data Medical System in Developing Countries. *Symmetry* **2018**, *10*, 152. [CrossRef]
26. Alharbi, S.; Rambely, A.S. Stability Analysis of Mathematical Model on the Effect of Modern Lifestyles Towards the Immune System. *J. Qual. Meas. Anal.* **2018**, *14*, 99–114.
27. Rosen, B.; Israeli, A.; Shortell, S. *Accountability and Responsibility in Health Care: Issues in Addressing an Emerging Global Challenge*; World Scientific: London, UK, 2012.
28. Cairns, J. *Cancer: Science and Society*; WH Freeman: New York, NY, USA, 1978.
29. Karacabey, K.; Ozdemir, N. The Effect of Nutritional Elements on the Immune System. *J. Obes. Weight Loss Ther.* **2012**, *2*, 152. [CrossRef]
30. World Cancer Research Fund; American Institute for Cancer Research. Diet, Nutrition, Physical Activity and Cancer: A Global Perspective. 2018. Available online: https://www.wcrf.org/sites/default/files/Summary-third-expert-report.pdf (accessed on 1 August 2018).
31. Rondanelli, M.; Faliva, M.A.; Miccono, A.; Naso, M.; Nichetti, M.; Riva, A.; Guerriero, F.; De Gregori, M.; Peroni, G.; Perna, S. Food pyramid for subjects with chronic pain: Foods and dietary constituents as anti-inflammatory and antioxidant agents. *Nutr. Res. Rev.* **2018**, *31*, 131–151. [CrossRef] [PubMed]
32. Achadi, E.; Ahuja, A.; Bendech, M.A.; Bhutta, Z.A.; De-Regil, L.M.; Fanzo, J.; Fracassi, P.; Grummer-Strawn, L.M.; Haddad, L.J.; Hawkes, C.; et al. *Global Nutrition Report: From Promise to Impact: Ending Malnutrition by 2030*; International Food Policy Research Institute: Washington, DC, USA, 2016.
33. Marwitz, S.E.; Woodie, L.N.; Blythe, S.N. Western-style diet induces insulin insensitivity and hyperactivity in adolescent male rats. *Physiol. Behav.* **2015**, *151*, 147–154. [CrossRef] [PubMed]

34. Sample, C.H.; Martin, A.A.; Jones, S.; Hargrave, S.L.; Davidson, T.L. Western-style diet impairs stimulus control by food deprivation state cues: Implications for obesogenic environments. *Appetite* **2015**, *93*, 13–23. [CrossRef]
35. Shaharudin, S.H.; Sulaiman, S.; Shahril, M.R.; Emran, N.A.; Akmal, S.N. Dietary changes among breast cancer patients in Malaysia. *Cancer Nurs.* **2013**, *36*, 131–138. [CrossRef] [PubMed]
36. Park, Y.; Subar, A.F.; Hollenbeck, A.; Schatzkin, A. Dietary fiber intake and mortality in the NIH-AARP diet and health study. *Arch. Intern. Med.* **2011**, *171*, 1061–1068. [CrossRef]
37. Tilg, H.; Moschen, A.R. Food, immunity, and the microbiome. *Gastroenterology* **2015**, *148*, 1107–1119. [CrossRef]
38. Uranga, J.A.; López-Miranda, V.; Lombo, F.; Abalo, R. Food, nutrients and nutraceuticals affecting the course of inflammatory bowel disease. *Pharmacol. Rep.* **2016**, *68*, 816–826. [CrossRef]
39. Newmark, H.L.; Yang, K.; Kurihara, N.; Fan, K.; Augenlicht, L.H.; Lipkin, M. Western-style diet-induced colonic tumors and their modulation by calcium and vitamin D in C57Bl/6 mice: A preclinical model for human sporadic colon cancer. *Carcinogenesis* **2008**, *30*, 88–92. [CrossRef] [PubMed]
40. Hatami, M.; Esmaeil Akbari, M.; Abdollahi, M.; Ajami, M.; Jamshidinaeini, Y.; Davoodi, S.H. The relationship between intake of macronutrients and vitamins involved in one carbon metabolism with breast cancer risk. *Tehran Univ. Med. J. TUMS Publ.* **2017**, *75*, 56–64.

© 2019 by the authors. Licensee MDPI, Basel, Switzerland. This article is an open access article distributed under the terms and conditions of the Creative Commons Attribution (CC BY) license (http://creativecommons.org/licenses/by/4.0/).

Article

Propagation of Blast Waves in a Non-Ideal Magnetogasdynamics

Astha Chauhan [1], Rajan Arora [1] and Mohd Junaid Siddiqui [2,*]

1. Department of Applied Science and Engineering, Indian Institute of Technology Roorkee, Roorkee 247667, India; asthaiitr1@gmail.com (A.C.); rajanfpt@iitr.ac.in (R.A.)
2. Department of Mathematics and Natural Sciences (DMNS), Prince Mohammad Bin Fahd University, Dhahran 34754, Saudi Arabia
* Correspondence: msiddiqui@pmu.edu.sa or kjunaidsiddiqui@gmail.com; Tel.: +966-544316033

Received: 6 March 2019; Accepted: 27 March 2019; Published: 1 April 2019

Abstract: Blast waves are generated when an area grows abruptly with a supersonic speed, as in explosions. This problem is quite interesting, as a large amount of energy is released in the process. In contrast to the situation of imploding shocks in ideal gas, where a vast literature is available on the effect of magnetic fields, very little is known about blast waves propagating in a magnetic field. As this problem is highly nonlinear, there are very few techniques that may provide even an approximate analytical solution. We have considered a problem on planar and radially symmetric blast waves to find an approximate solution analytically using Sakurai's technique. A magnetic field has been taken in the transverse direction. Gas particles are supposed to be propagating orthogonally to the magnetic field in a non-deal medium. We have further assumed that specific conductance of the medium is infinite. Using Sakurai's approach, we have constructed the solution in a power series of $(C/U)^2$, where C is the velocity of sound in an ideal gas and U is the velocity of shock front. A comparison of obtained results in the absence of a magnetic field within the published work of Sakurai has been made to generate the confidence in our results. Our results match well with the results reported by Sakurai for gas dynamics. The flow variables are computed behind the leading shock and are shown graphically. It is very interesting that the solution of the problem is obtained in closed form.

Keywords: blast waves; non-ideal gas; Rankine–Hugoniot conditions; magnetogasdynamics

MSC: 35L67; 58J45

1. Introduction

The study of propagation of strong shock waves has always been given undivided attention by various research groups in the field of science. In particular, it is useful in nuclear science, plasma physics, geophysics, astrophysics, explosions, etc. Shocks are ubiquitary and quite useful in the generation of energy. Radially converging shocks are used to create a hot spot at the center in Inertial Confinement Fusion (ICF). The heat generated in the process is used to activate nuclear fission. After World War II, it becomes extremely important to understand explosion dynamics. Motivated by this, Sedov [1] first coined the idea of a similarity solution for the point explosion problem in an ideal medium. Soon, this solution became famous as the "Sedov Similarity Solution" among researchers due to its importance in blast wave theory. Taylor [2] found the analytical first approximate solution for the problem. These results of Sedov and Taylor showed a way to estimate the effects of nuclear or supernova explosions. Later, Sakurai [3,4] used this first approximate solution to find self-similar solutions to the problem. Murata [5] and Donato [6] obtained the exact solutions to blast wave problems in gas dynamics. Murata [5] assumed that change in density ahead of shock front is governed by the exponent of distance from the point (or axis) of ignition. The density ahead of the leading shock was assumed to vary as the power of the radial distance from the center of explosion in his study. Book [7] used Sedov formulas to find similarity solution of the point explosion problem. Lalicata and Torrisi [8] used the similarity method for reacting flow. Propagation of shock waves generated by sudden explosions within the presence of a magnetic field is very important because this phenomenon is well marked in the disassembled nature of the absorption of energy. Anisimov and Spiner [9] and Lerche [10] presented the theory of blast waves in magnetogasdynamics. A first comprehensive study on the effect of magnetic field on the exploding shocks was presented by Lerche [11] in 1979. There is an abrupt change in the entropy across an exploding shock and it releases a very high energy in the process. It is well known that after the explosion, the front of the blast wave is circumscribed by the surfaces of the shocks and propagates with decreasing velocity. Furthermore, a very high energy is released in the process. It raises the temperature of the surrounding. Therefore, the assumption of an ideal medium is not valid anymore, and one must consider the effect of non-deal medium on the blast waves. The gas particles are ionized due to the presence of high temperature at the center. These ionized gas particles generate the magnetic field. Therefore, the inclusion of a magnetic field on the study of blast waves is essential for capturing the physics of the process well. It has many applications in oceanography, astrophysics, aerodynamics, and atmospheric sciences.

Many research groups have worked afterwards on the topic to better understand the dynamics of shock waves in a magnetic field. Among the recent research on the topic, we wish to mention the work of Arora et al. [12], Siddiqui et al. [13], Singh et al. [14,15], and Pandey et al. [16]. Menon and Sharma [17] studied the flattening and steepening of the characteristics wave fronts in an ideal medium with magnetic field. Arora et al. [18] found a similar solution to the propagation of shocks in a non-ideal medium. Relaxation effects were also included in the study. Later, Siddiqui et al. [19] used asymptotic expansion of flow variables to find the solution to nonlinear waves far from the origin. The relaxation of gas particles has been taken into account in the study. Evolutionary behavior of weak shocks in real gas is presented by Arora and Siddiqui [20]. Despite there being vast literature available on imploding shocks in real gas, the behavior of shock front after ignition is still an open problem. In the present work, we have analyzed the flow variables after a blast in ignition by using the Sakurai analytical approach. The medium is considered to be real gases. Effects of a magnetic field have also been taken into account.

This paper is summarized as follows: Section 1 contains a brief introduction to the topic and historical background of earlier studies. Section 2 presents the fundamental equations governing the conservation laws. Rankine–Hugoniot (RH) jump conditions are also presented in this section. In Section 3, we introduce the new independent variables and transform the fundamental equations in the form of non-dimensional functions using the similarity analysis. In Section 4, power series solutions in terms of $(C/U)^2$ have been presented for the problem. In Section 5, the first approximation solutions are obtained which correspond to Taylor's series solution. In Section 6, a brief conclusion is given about the whole study of this paper. Based on the study, we have concluded that the density and magnetic pressure of the particle decreases behind the leading shock. Charged particles are transported away by the magnetic field, and this gives an increase in the velocity of blast wind.

2. Fundamental Equations

The fundamental equations which govern unsteady planar ($m = 0$) or cylindrically ($m = 1$) symmetric flow in a non-ideal gas in the presence of transverse magnetic field can be expressed as [12,21]

$$\rho_t + \rho u_x + u\rho_x + \frac{m\rho u}{x} = 0, \quad (1)$$

$$u_t + uu_x + \frac{1}{\rho}(p_x + h_x) = 0, \quad (2)$$

$$p_t + u p_x + a^2 \rho \left(u_x + \frac{mu}{x} \right) = 0, \qquad (3)$$

$$h_t + u h_x + 2h \left(u_x + \frac{mu}{x} \right) = 0. \qquad (4)$$

Here ρ, u, p and $h = \frac{\mu H^2}{2}$ denote the density, velocity, pressure, and magnetic pressure, respectively. μ denotes the magnetic permeability and H being used for transverse magnetic field. $a^2 = \frac{\gamma p}{\rho(1-b\rho)}$ is speed of sound in real gases. γ is the ratio the specific heats at constant pressure and volume of the gas. t and x represents the time and space variables, respectively. $m = 0$ denotes to planar symmetry while $m = 1$ denotes cylindrical symmetry of the flow. It is assumed that electrical resistivity of the medium is zero and magnetic field is orthogonal to the radial motion of gas particles.

This system of Equations (1)–(4) is supplemented with the equation of state for real gas as follows

$$p = \frac{\rho R T}{(1 - b\rho)}. \qquad (5)$$

R, T represent the gas constant and temperature, respectively.

The position of leading shock at any time t is given by $\tilde{R} = \tilde{R}(t)$. Therefore, the velocity of shock front is ($\frac{d\tilde{R}}{dt} = C$). The conditions for flow variables ahead of shock are characterized by

$$\rho = \rho_0(x), \quad u = 0, \quad p = p_0, \quad h = h_0(x). \qquad (6)$$

RH conditions at the leading shock ($x = \tilde{R}(t)$) are obtained by the conservation of laws that can be expressed in simplified form as follows [22]

$$(\rho)_{x=R} = \frac{(\gamma+1)}{(\gamma-1)} \rho_0 \left[1 - \frac{2\alpha}{\gamma-1} - \frac{2}{\gamma-1} \left(\frac{C}{U} \right)^2 \right], \qquad (7)$$

$$(u)_{x=R} = \frac{2}{(\gamma-1)} S \left[1 - \alpha - \left(\frac{V}{S} \right)^2 \right], \qquad (8)$$

$$(p)_{x=R} = \frac{2\rho_0 U^2}{\gamma+1} \left[1 - \alpha - \frac{\gamma-1}{2\gamma} \left(\frac{C}{U} \right)^2 \right] - \frac{1}{2} \left(\frac{\gamma+1}{\gamma-1} \right)^2 C_0 \rho_0 U^2 \left[1 - \frac{4\alpha}{\gamma-1} - \frac{4}{\gamma-1} \left(\frac{C}{U} \right)^2 \right], \qquad (9)$$

$$(h)_{x=R} = \frac{1}{2} \left(\frac{\gamma+1}{\gamma-1} \right)^2 C_0 \rho_0 U^2 \left[1 - \frac{4\alpha}{\gamma-1} - \frac{4}{\gamma-1} \left(\frac{C}{U} \right)^2 \right], \qquad (10)$$

where $C_0 = \frac{2h_0}{\rho_0 U^2}$ is the shock cowling number, $C^2 = \frac{\gamma p_0}{\rho_0}$ is the velocity of shock and $\alpha = (\gamma - 1) b \rho_0$. The initial density ρ_0 is assumed to follow the power law given as

$$\rho_0 = \rho_c x^{-\delta}, \qquad (11)$$

where x is the perpendicular distance of the point on leading shock from the point of explosion. ρ_c is the constant density and δ is an exponent. Using the conservation of total energy, we get

$$E = \int_0^{\tilde{R}} \left[\frac{1}{2} u^2 + \frac{(1-b\rho)}{(\gamma-1)} \left(\frac{p}{\rho} - \frac{p_0}{\rho_0} \right) + \left(\frac{h}{\rho} - \frac{h_0}{\rho_0} \right) \right] \rho x^m dx, \quad m = 0, 1, \qquad (12)$$

where E denotes the surface energy carried by blast waves per unit of area. We obtain the following relation by the Lagrangian equation of continuity:

$$\int_0^{\tilde{R}} \frac{\rho}{\rho_0} x^m dx = \frac{\tilde{R}^{m+1}}{m+1}. \qquad (13)$$

We use the value of $\int_0^{\tilde{R}} \frac{\rho}{\rho_0} x^m dx$ from Equation (13) in Equation (12) to get the following expression

$$E = \int_0^{\tilde{R}} \left[\frac{1}{2}\rho u^2 + \frac{(1-b\rho)}{(\gamma-1)}p + h\right] x^m dx - \left\{\frac{(1-b\rho)p_0}{\gamma-1} + h_0\right\} \frac{\tilde{R}^{m+1}}{m+1}. \tag{14}$$

3. Similarity Transformation of Fundamental Equations

As the process is self-similar, $x = R(t)$ and $U = \frac{d\tilde{R}}{dt}$. It gives us the characteristic scale. We, therefore, introduce the following new independent non-dimensional variables in place of x and S:

$$x/\tilde{R} = r, \quad (C/U)^2 = s. \tag{15}$$

$x/\tilde{R} = r$ is known as similarity variable. Non-dimensional quantities u/U, p/p_0, ρ/ρ_0 and $h/(p_0(U/C)^2)$ are assumed to be equal to F, G, Π, and H, respectively which are consistent with Equations (1)–(4). We, therefore, can take the similarity transformations as:

$$u = UF(r,s), \tag{16}$$

$$p = p_0(U/C)^2 G(r,s), \tag{17}$$

$$\rho = \rho_0 \Pi(r,s), \tag{18}$$

$$h = p_0(U/C)^2 H(r,s) = p_0 H(r,s)/s. \tag{19}$$

Using the system of Equations (16)–(19), we obtain

$$\frac{\partial}{\partial x} = \frac{1}{\tilde{R}} \frac{\partial}{\partial r}, \tag{20}$$

$$\frac{D}{Dt} = \frac{U}{\tilde{R}}\left[(F-r)\frac{\partial}{\partial r} + \lambda s \frac{\partial}{\partial s}\right], \tag{21}$$

where $\lambda = \lambda(s) = \tilde{R}(ds/d\tilde{R})/s$. Substituting (16)–(19) into the fundamental Equations (1)–(4), we obtain

$$(F-r)\Pi_r + \lambda s \Pi_s + \Pi\left(F_r + \frac{mF}{r}\right) = 0, \tag{22}$$

$$\Pi\left(-\frac{\lambda}{2}F + (F-r)F_r + \lambda s F_s\right) + \frac{1}{\gamma}(G_r + H_r) = 0, \tag{23}$$

$$-\lambda G + (F-r)G_r + \lambda s G_s + \frac{\gamma G}{(1-\bar{b}\Pi)}\left(F_r + \frac{mF}{r}\right) = 0, \tag{24}$$

$$-\lambda H + (F-r)H_r + \lambda s H_s + 2H\left(F_r + \frac{mF}{r}\right) = 0, \tag{25}$$

where $\bar{b} = b\rho_0$. Using Equation (14), we obtain

$$s\left(\frac{\tilde{R}_0}{\tilde{R}}\right)^{m+1} = \int_0^1 \left(\frac{1}{2}\gamma \Pi F^2 + \frac{(1-\bar{b}\Pi)G}{(\gamma-1)} + H\right) r^m dr - \frac{(1-\bar{b}\Pi)s}{(\gamma-1)(m+1)} - \frac{\gamma C_0}{2(m+1)}, \tag{26}$$

where

$$\tilde{R}_0 = \left(\frac{E}{p_0}\right)^{1/(m+1)}. \tag{27}$$

Using (16)–(19), RH conditions become

$$\Pi(1,s) = \frac{(\gamma+1)}{(\gamma-1)}\left[1 - \frac{2\alpha}{\gamma-1} - \frac{2}{\gamma-1}s\right], \quad (28)$$

$$F(1,s) = \frac{2}{\gamma+1}(1-\alpha-s), \quad (29)$$

$$G(1,s) = \frac{2\gamma}{\gamma+1}\left[1 - \alpha - \frac{(\gamma-1)s}{2\gamma}\right] - \frac{1}{2}\left(\frac{\gamma+1}{\gamma-1}\right)^2 C_0\gamma\left[1 - \frac{4\alpha}{\gamma-1} - \frac{4s}{\gamma-1}\right], \quad (30)$$

$$H(1,s) = \frac{\gamma}{2}C_0\left(\frac{\gamma+1}{\gamma-1}\right)^2\left[1 - \frac{4\alpha}{\gamma-1} - \frac{4s}{\gamma-1}\right]. \quad (31)$$

Now differentiating Equation (26) regarding s, we obtain the expression for λ as

$$\lambda = \frac{J(m+1) - \frac{(1-\bar{b}\Pi)s}{\gamma-1} - \frac{\gamma C_0}{2}}{J - s\frac{dJ}{ds} - \frac{\gamma C_0}{2(m+1)}}. \quad (32)$$

Here

$$J = \int_0^1 \left(\frac{1}{2}\gamma\Pi F^2 + \frac{(1-\bar{b}\Pi)G}{(\gamma-1)} + H\right)r^m dr. \quad (33)$$

The self-similar process of the explosion has clearly explained the power rule particularly in Equation (27).

4. Construction of Solutions in Power Series of s

As we are seeking solution for strong shocks, the velocity of leading shock front S is very large than the velocity of sound waves V in ideal gas. The quantity $s = (V/S)^2$ is very small for strong shocks. We, therefore, expand the non-dimensional quantities F, G, Π and H in power series $s = (V/S)^2$ as follows

$$\begin{aligned}
F &= F^{(0)} + sF^{(1)} + s^2F^{(2)} + s^3F^{(3)} + ..., \\
G &= G^{(0)} + sG^{(1)} + s^2G^{(2)} + s^3G^{(3)} + ..., \\
\Pi &= \Pi^{(0)} + s\Pi^{(1)} + s^2\Pi^{(2)} + s^3\Pi^{(3)} + ..., \\
H &= H^{(0)} + sH^{(1)} + s^2H^{(2)} + s^3H^{(3)} + ...
\end{aligned} \quad (34)$$

Here $F^{(i)}$, $G^{(i)}$, $\Pi^{(i)}$ and $H^{(i)}$ ($i = 0, 1, 2, ...$) are either constants or functions of r. We use the power series expansions from Equation (34) in Equation (33) to obtain the value of J as

$$J = J_0(1 + \sigma_1 s + \sigma_2 s^2 + ...), \quad (35)$$

where

$$\begin{aligned}
J_0 &= \int_0^1 \left[\frac{\gamma}{2}\Pi^{(0)}(F^{(0)})^2 + \frac{(1-\bar{b}\Pi^{(0)})}{\gamma-1}G^{(0)} + H^{(0)}\right]r^m dr, \\
\sigma_1 J_0 &= \int_0^1 \left[\frac{\gamma}{2}\Pi^{(1)}(F^{(0)})^2 + \gamma\Pi^{(0)}F^{(0)}F^{(1)} + \frac{1}{\gamma-1}(G^{(1)} - \bar{b}\Pi^{(0)}G^{(1)} - \bar{b}\Pi^{(1)}G^{(0)}) + H^{(1)}\right]r^m dr, \\
\sigma_2 J_0 &= \int_0^1 \left[\frac{\gamma}{2}\Pi^{(2)}(F^{(0)})^2 + \gamma\Pi^{(0)}F^{(0)}F^{(2)} + \frac{\gamma}{2}\Pi^{(0)}(F^{(1)})^2 + 2\Pi^{(1)}F^{(1)}F^{(0)} + H^{(2)}\right]r^m dr \\
&+ \int_0^1 \frac{1}{\gamma-1}[G^{(2)} - \bar{b}\Pi^{(2)}G^{(0)} - \bar{b}\Pi^{(1)}G^{(1)} - \bar{b}\Pi^{(0)}G^{(2)}]r^m dr, \\
&...
\end{aligned} \quad (36)$$

Using (35) in (26), we obtain

$$s\left(\frac{\tilde{R}_0}{\tilde{R}}\right)^{m+1} = J_0\left[\left(1 - \frac{\gamma C_0}{2(m+1)J_0}\right) + \left(\sigma_1 - \frac{1}{(m+1)(\gamma-1)J_0}\right)s + \sigma_2 s^2 + \ldots\right]. \quad (37)$$

In view of (15), Equation (37) becomes

$$\left(\frac{V}{S}\right)^2 \left(\frac{\tilde{R}_0}{\tilde{R}}\right)^{m+1} = J_0\left[\left(1 - \frac{\gamma C_0}{2(m+1)J_0}\right) + \left(\sigma_1 - \frac{1}{(m+1)(\gamma-1)J_0}\right)\left(\frac{V}{S}\right)^2 + \sigma_2\left(\frac{V}{S}\right)^4 + \ldots\right]. \quad (38)$$

This equation provides a relation between shock velocity and its position at time t. We expand λ by using Equations (32) and (35) as follows:

$$\lambda = (m+1)[1 + \sigma'_1 s + 2\sigma'_2 s^2 + \ldots], \quad (39)$$

where

$$\sigma'_1 = \frac{\sigma_1 - \frac{1}{(m+1)(\gamma-1)J_0}}{1 - \frac{\gamma C_0}{2(m+1)(\gamma-1)J_0}},$$

$$\sigma'_2 = \frac{\sigma_2}{1 - \frac{\gamma C_0}{2(m+1)J_0}},$$

$$\sigma'_3 = \frac{\sigma_3}{1 - \frac{\gamma C_0}{2(m+1)J_0}}.$$

For simplification, let us consider $\lambda_1 = \sigma'_1, \lambda_2 = 2\sigma'_2, \lambda_3 = 3\sigma'_3,\ldots$, then the Equation (39) becomes

$$\lambda = (m+1)[1 + \lambda_1 s + \lambda_2 s^2 + \ldots]. \quad (40)$$

We use the relations from Equations (34) and (40) in Equations (22)–(25). Comparing the likewise powers of s on both sides of an equation, we get relations in terms of Ordinary Differential Equations (ODEs). Comparing the terms free from s, we get

$$(F^{(0)} - r)\Pi_r^{(0)} + \Pi^{(0)}\left(F_r^{(0)} + \frac{mF^{(0)}}{r}\right) = 0, \quad (41)$$

$$(F^{(0)} - r)\Pi^{(0)}F_r^{(0)} + \frac{1}{\gamma}(G_x^{(0)} + H_x^{(0)}) = \frac{(m+1)F^{(0)}\Pi^{(0)}}{2}, \quad (42)$$

$$-(m+1)G^{(0)} + (F^{(0)} - r)G_r^{(0)} + \frac{\gamma G^{(0)}}{(1 - b\Pi^{(0)})}\left(F_r^{(0)} + \frac{mF^{(0)}}{r}\right) = 0, \quad (43)$$

$$-(m+1)H^{(0)} + (F^{(0)} - r)H_r^{(0)} + 2H^{(0)}\left(F_r^{(0)} + \frac{mF^{(0)}}{r}\right) = 0. \quad (44)$$

Equating the power s, we get

$$(F^{(0)} - r)\Pi_r^{(1)} + F^{(1)}\Pi_r^{(0)} + (m+1)\Pi^{(1)} + \Pi^{(0)}\left(F_r^{(1)} + \frac{mF^{(1)}}{r}\right)$$
$$+ \Pi^{(1)}\left(F_r^{(0)} + \frac{mF^{(0)}}{r}\right) = 0, \qquad (45)$$

$$-\frac{\Pi^{(0)}}{2}(m+1)[F^{(1)} + \lambda_1 F^{(0)}] - \frac{(m+1)\Pi^{(1)}F^{(0)}}{2} + \Pi^{(0)}[(F^{(0)} - r)F_r^{(1)} + F^{(1)}F_r^{(0)}]$$
$$+ \Pi^{(1)}(F^{(0)} - r)F_r^{(0)} + (m+1)\Pi^{(0)}F^{(1)} + \frac{1}{\gamma}[G_r^{(1)} + H_r^{(1)}] = 0, \qquad (46)$$

$$-\lambda G^{(1)} + F^{(1)}G_r^{(0)} + \lambda_1(G_r^{(0)} + G^{(1)}) + F^{(0)}G_r^{(1)} + [1 + b\Pi^{(1)}]G^{(0)}$$
$$+ \{b\Pi^{(0)} + (b\Pi^{(0)})^2 + ...\}G^{(1)}][F^{(0)} + \frac{mF^{(0)}}{r}] = 0, \qquad (47)$$

$$-(m+1)\lambda_1 H^{(0)} + (F^{(0)} - r)H_r^{(1)} + F^{(1)}H_r^{(0)} + 2H^{(0)}\left(F_r^{(1)} + \frac{mF^{(1)}}{r}\right)$$
$$+ 2H^{(1)}\left(F_r^{(0)} + \frac{mF^{(0)}}{r}\right) = 0, \qquad (48)$$

From Equations (28)–(31) and (34), we have

$$F^{(0)}(1) = \frac{2}{\gamma+1}(1-\alpha), \quad G^{(0)}(1) = \frac{2\gamma}{\gamma+1}(1-\alpha) - \left(\frac{\gamma+1}{\gamma-1}\right)^2 \frac{\gamma C_0}{2}\left(1 - \frac{4\alpha}{\gamma-1}\right),$$
$$\Pi^{(0)}(1) = \frac{\gamma+1}{\gamma-1}\left(1 - \frac{2\alpha}{\gamma-1}\right), \quad H^{(0)}(1) = \left(\frac{\gamma+1}{\gamma-1}\right)^2 \frac{\gamma C_0}{2}\left(1 - \frac{4\alpha}{\gamma-1}\right). \qquad (49)$$

$$F^{(1)}(1) = -\frac{2}{\gamma+1}, \quad G^{(1)}(1) = -\frac{\gamma-1}{\gamma+1} + 2\gamma C_0 \frac{(\gamma+1)^2}{(\gamma-1)^3},$$
$$\Pi^{(1)}(1) = -\frac{2(\gamma+1)}{(\gamma-1)}, \quad H^{(1)}(1) = -2\gamma C_0 \frac{(\gamma+1)^2}{(\gamma-1)^3}. \qquad (50)$$

To get the first approximate solution, we determine $F^{(0)}$, $G^{(0)}$, $\Pi^{(0)}$ and $H^{(0)}$ from the system of nonlinear ODEs (41)–(44) with the boundary conditions given in Equation (49). Finally, these values are used in Equation $(36)_1$ to get approximate solution as follows

$$u = SF^{(0)}(r), \quad p = \rho_0(S/V)^2 G^{(0)}(r),$$
$$\rho = \rho_0 \Pi^{(0)}(r), \quad h = \rho_0(S/V)^2 H^{(0)}(r). \qquad (51)$$

To get the second approximation, we need to determine the values of $F^{(1)}$, $G^{(1)}$, $\Pi^{(1)}$ and $H^{(1)}$ in Equation (45). $F^{(0)}$, $G^{(0)}$, $\Pi^{(0)}$ and $H^{(0)}$ are used from the first approximation. $F^{(1)}$, $G^{(1)}$, $\Pi^{(1)}$ and $H^{(1)}$ contain λ_1. We use these values in terms of λ_1 in Equation $(36)_2$ to finally obtain the value of λ_1. The obtained value of λ_1 is used to get the second approximate solution. The other steps involve the repetition of the above process to get higher order approximate solutions of the problem.

5. The First Approximation

The Equations (41)–(44) can be written in the following forms:

$$\frac{\Pi_r^{(0)}}{\Pi^{(0)}} = \left(F_r^{(0)} + \frac{mF^{(0)}}{r}\right) / (r - F^{(0)}), \qquad (52)$$

$$(F^{(0)} - r)\Pi^{(0)} F_r^{(0)} + \frac{1}{\gamma}(G_r^{(0)} + H_r^{(0)}) = \frac{(m+1)F^{(0)}\Pi^{(0)}}{2}, \qquad (53)$$

$$\frac{G_r^{(0)}}{G^{(0)}} = \left[-(m+1) + \frac{\gamma}{(1-\bar{b}\Pi^{(0)})}\left(F_r^{(0)} + \frac{mF^{(0)}}{r}\right)\right]/(r - F^{(0)}), \qquad (54)$$

$$\frac{H_r^{(0)}}{H^{(0)}} = \left(2F_r^{(0)} + \frac{2mF^{(0)}}{r} - m - 1\right)/(r - F^{(0)}). \qquad (55)$$

We substitute the values $G_r^{(0)}$ and $H_r^{(0)}$ from Equations (54) and (55) in Equation (53) to get $F_r^{(0)}$ in the following form

$$F_r^{(0)} = \frac{\left[\frac{m+1}{\gamma} - \frac{mF^{(0)}}{(1-\bar{b}\Pi^{(0)})r} + \left(\frac{m+1}{\gamma} - \frac{2mF^{(0)}}{\gamma r}\right)\frac{H^{(0)}}{G^{(0)}} + \frac{(m+1)F^{(0)}\Pi^{(0)}(r-F^{(0)})}{2G^{(0)}}\right]}{\left[\frac{2H^{(0)}}{\gamma G^{(0)}} + \frac{1}{(1-\bar{b}\Pi^{(0)})} - \frac{\Pi^{(0)}(r-F^{(0)})}{G^{(0)}}\right]}. \qquad (56)$$

Taylor [23] performed similarity transformation and presented the approximate solution for intense explosion. This approximation has been used by Sakurai [3] to produce analytical solution in gas dynamics. We, therefore, has assumed the first approximation $F^{(0)}$ as given in the work of Taylor [23]

$$F^{(0)}(r) = \frac{r}{\gamma} + Ar^n. \qquad (57)$$

We get the value of A in the following form using Equations (49) and (57),

$$A = \frac{\gamma(1 - 2\alpha) - 1}{\gamma(\gamma + 1)}. \qquad (58)$$

We use Equations (56)–(58) to get n. After determining the values of A and n, we integrate Equations (52)–(55) with boundary conditions (49) to obtain

$$\Pi^{(0)}(r) = \frac{\gamma+1}{\gamma-1}\left(1 - \frac{2\alpha}{\gamma-1}\right)\left[\frac{\gamma}{\gamma+1-r^{n-1}}\right]^{\left(\frac{(n+m)(\gamma(1-2\alpha)-1)}{(n-1)(\gamma-1)} + \frac{m+1}{(n-1)(\gamma-1)}\right)} r^{\left(\frac{m+1}{\gamma-1}\right)}, \qquad (59)$$

$$G^{(0)}(r) = \left\{\frac{2\gamma}{\gamma+1}(1-\alpha) - \left(\frac{\gamma+1}{\gamma-1}\right)^2 \frac{\gamma C_0}{2}\left(1 - \frac{4\alpha}{\gamma-1}\right)\right\}\left[\frac{\gamma}{\gamma+1-r^{n-1}}\right]^{(B)} r^{\frac{(\gamma+1)(m+1)\bar{b}(1-\frac{2\alpha}{\gamma-1})}{(\gamma-1)^2[1-\bar{b}\frac{\gamma+1}{\gamma-1}(1-\frac{2\alpha}{\gamma-1})]}}, \qquad (60)$$

where

$$B = \frac{\gamma(m+n)(\gamma(1-2\alpha) - 1) + \gamma(m+1)\bar{b}(\gamma+1)(1 - \frac{2\alpha}{\gamma-1})}{(n-1)(\gamma-1)\{1 - \bar{b}(\frac{\gamma+1}{\gamma-1})(1 - \frac{2\alpha}{\gamma-1})\}}.$$

and

$$H^{(0)}(r) = \frac{\gamma C_0}{2}\left(\frac{\gamma+1}{\gamma-1}\right)^2\left(1 - \frac{4\alpha}{\gamma-1}\right)\left[\frac{\gamma}{\gamma+1-r^{n-1}}\right]^{\left(\frac{2(m+n)(\gamma(1-2\alpha)-1)+(1+m)(2-\gamma)}{(n-1)(\gamma-1)}\right)} r^{\frac{(m+1)(2-\gamma)}{(\gamma-1)}}. \qquad (61)$$

Dimensionless quantities $F^{(0)}$, $\Pi^{(0)}$ and $G^{(0)}$ are computed in Tables 1 and 2 for $(m = 0, 1)$ in ideal gas (without magnetic field i.e., $C_0 = 0$). A comparison of the obtained results is presented with the published work of Sakurai [4] in gas dynamics. Schematic of dimensionless flow variables are depicted in Figures 1–4.

Table 1. $F^{(0)}$, $\Pi^{(0)}$ and $G^{(0)}$ for $m = 0$, $\gamma = 1.4$ and $C_0 = 0$.

x	α	$F^{(0)}$ Sakurai [3]	$\Pi^{(0)}$ Sakurai [3]	$G^{(0)}$ Sakurai [3]	$F^{(0)}$	$\Pi^{(0)}$	$G^{(0)}$
	0.0	0.8333	6.000	1.167	0.8333	6.000	1.167
1.0	0.015	-	-	-	0.8208	5.550	1.167
	0.025	-	-	-	0.8125	5.250	1.167
	0.0	0.7170	3.096	0.815	0.7251	3.096	0.815
0.9	0.015	-	-	-	0.7199	2.899	0.835
	0.025	-	-	-	0.7159	2.767	0.848
	0.0	0.6151	1.785	0.647	0.6259	1.785	0.647
0.8	0.015	-	-	-	0.6251	1.660	0.664
	0.025	-	-	-	0.6240	1.577	0.677
	0.0	0.5239	1.080	0.556	0.5341	1.080	0.556
0.7	0.015	-	-	-	0.5356	0.990	0.567
	0.025	-	-	-	0.5362	0.931	0.575
	0.0	0.4405	0.658	0.504	0.4484	0.658	0.504
0.6	0.015	-	-	-	0.4507	0.593	0.508
	0.025	-	-	-	0.4521	0.551	0.515
	0.0	0.3624	0.389	0.473	0.3676	0.389	0.473
0.5	0.015	-	-	-	0.3698	0.345	0.471
	0.025	-	-	-	0.3712	0.317	0.471
	0.0	0.2877	0.214	0.456	0.2877	0.214	0.456
0.4	0.015	-	-	-	0.2921	0.187	0.450
	0.025	-	-	-	0.2933	0.170	0.446
	0.0	0.2148	0.102	0.447	0.2160	0.102	0.447
0.3	0.015	-	-	-	0.2169	0.088	0.437
	0.025	-	-	-	0.2177	0.079	0.431
	0.0	0.1430	0.037	0.443	0.1432	0.037	0.443
0.2	0.015	-	-	-	0.1436	0.0316	0.432
	0.025	-	-	-	0.1439	0.0282	0.423
	0.0	0.0714	0.006	0.442	0.0714	0.006	0.442
0.1	0.015	-	-	-	0.0715	0.005	0.429
	0.025	-	-	-	0.0716	0.005	0.419
	0.0	0.0000	0.000	0.442	0.0000	0.000	0.442
0.0	0.015	-	-	-	0.0000	0.000	0.429
	0.025	-	-	-	0.0000	0.000	0.419

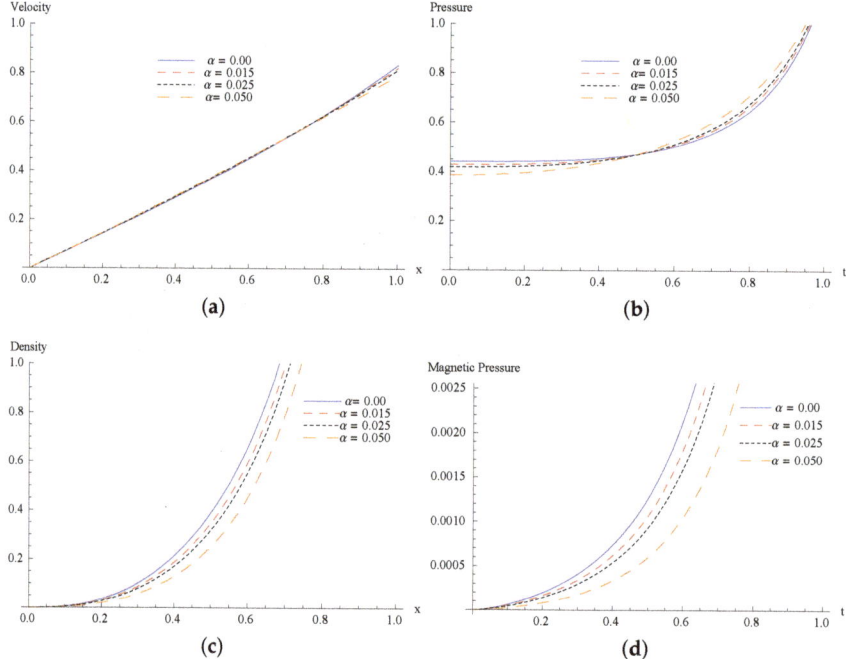

Figure 1. Schematic of non-dimensional velocity (**a**) $F^{(0)}$, (**b**) pressure $G^{(0)}$, (**c**) density $\Pi^{(0)}$ and (**d**) magnetic pressure $H^{(0)}$ for $\gamma = 1.4$, $C_0 = 0$, $m = 0$ and $\alpha = 0.0, 0.015, 0.025, 0.05$.

Table 2. $F^{(0)}$, $\Pi^{(0)}$ and $G^{(0)}$ for $m = 1$, $\gamma = 1.4$ and $C_0 = 0$.

x	α	$F^{(0)}$ Sakurai [3]	$\Pi^{(0)}$ Sakurai [3]	$G^{(0)}$ Sakurai [3]	$F^{(0)}$	$\Pi^{(0)}$	$G^{(0)}$
	0.0	0.8333	6.000	1.167	0.8333	6.000	1.167
1.0	0.015	-	-	-	0.8208	5.550	1.167
	0.025	-	-	-	0.8125	5.250	1.167
	0.0	0.7008	1.898	0.685	0.7072	1.898	0.684
0.9	0.015	-	-	-	0.7196	1.767	0.767
	0.025	-	-	-	0.7314	1.678	0.874
	0.0	0.5973	0.783	0.531	0.6038	0.783	0.531
0.8	0.015	-	-	-	0.6246	0.713	0.577
	0.025	-	-	-	0.6503	0.668	0.667
	0.0	0.5104	0.347	0.468	0.5141	0.346	0.468
0.7	0.015	-	-	-	0.5351	0.309	0.474
	0.025	-	-	-	0.5691	0.284	0.517
	0.0	0.4322	0.153	0.441	0.4346	0.150	0.441
0.6	0.015	-	-	-	0.4503	0.131	0.414
	0.025	-	-	-	0.4880	0.119	0.406
	0.0	0.3582	0.058	0.429	0.3592	0.0582	0.429
0.5	0.015	-	-	-	0.3695	0.050	0.378
	0.025	-	-	-	0.4077	0.045	0.323
	0.0	0.2859	0.019	0.425	0.2863	0.019	0.425
0.4	0.015	-	-	-	0.2919	0.016	0.357
	0.025	-	-	-	0.3255	0.014	0.260
	0.0	0.2140	0.005	0.424	0.2144	0.004	0.424
0.3	0.015	-	-	-	0.2168	0.004	0.346
	0.025	-	-	-	0.2443	0.003	0.211
	0.0	0.1429	0.001	0.424	0.1429	0.001	0.423
0.2	0.015	-	-	-	0.1437	0.001	0.340
	0.025	-	-	-	0.1630	0.000	0.173
	0.0	0.0714	0.000	0.424	0.0714	0.000	0.423
0.1	0.015	-	-	-	0.0715	0.000	0.338
	0.025	-	-	-	0.0815	0.000	0.143
	0.0	0.0000	0.000	0.424	0.0000	0.000	0.423
0.0	0.015	-	-	-	0.0000	0.000	0.338
	0.025	-	-	-	0.0000	0.000	0.118

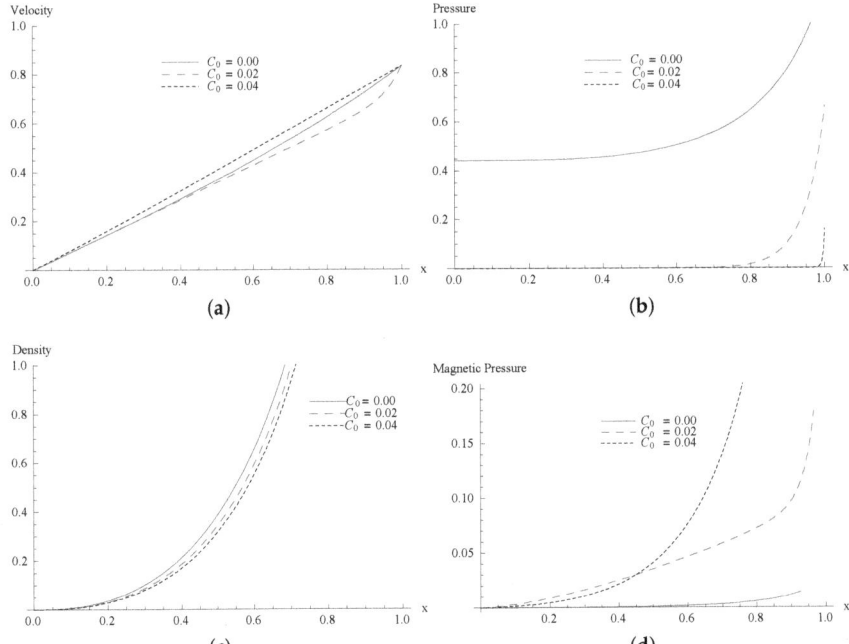

Figure 2. Schematic of non-dimensional (**a**) velocity $F^{(0)}$, (**b**) pressure $G^{(0)}$, (**c**) density $\Pi^{(0)}$ and (**d**) magnetic pressure $H^{(0)}$ for $\gamma = 1.4$, $\alpha = 0$, $m = 0$ and $C_0 = 0.00, 0.02, 0.04$.

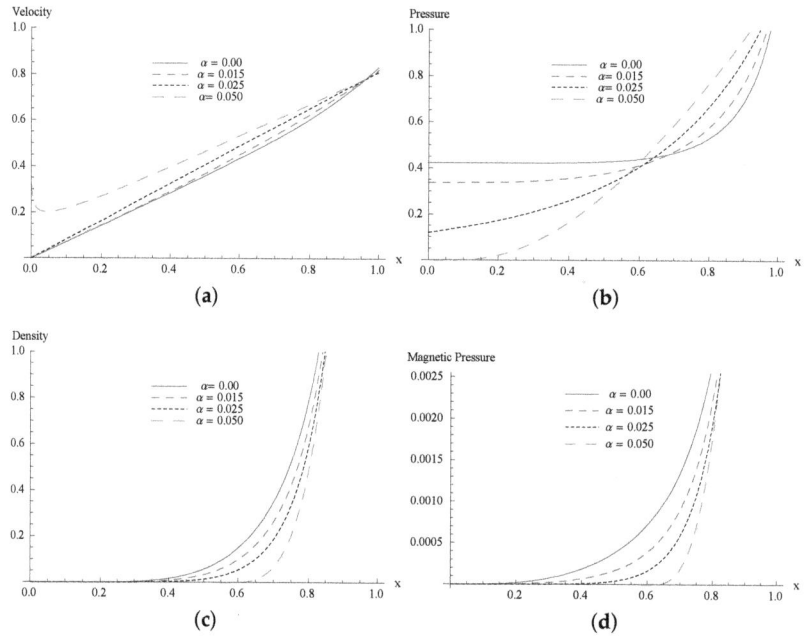

Figure 3. Schematic of non-dimensional (**a**) velocity $F^{(0)}$, (**b**) pressure $G^{(0)}$, (**c**) density $\Pi^{(0)}$ and (**d**) magnetic pressure $H^{(0)}$ for $\gamma = 1.4$, $C_0 = 0$, $m = 1$ and $\alpha = 0.0, 0.015, 0.025, 0.05$.

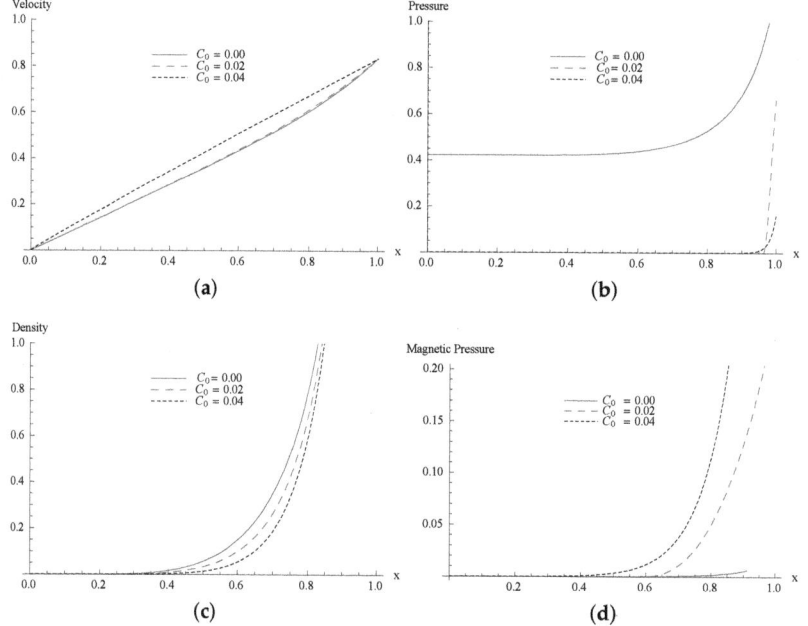

Figure 4. Schematic of non-dimensional (**a**) velocity $F^{(0)}$, (**b**) pressure $G^{(0)}$, (**c**) density $\Pi^{(0)}$ and (**d**) magnetic pressure $H^{(0)}$ for $\gamma = 1.4$, $\alpha = 0$, $m = 1$ and $C_0 = 0.00, 0.02, 0.04$.

6. Conclusions

In the present work, we have successfully used the Sakurai [4] approach to find the first-order approximate analytical solution of the planar and cylindrical symmetric flow under the presence of transverse magnetic field in a non-ideal medium. However, Sakurai [4] has obtained a solution in gas dynamics by using this method; there is no study afterwards that provides approximate analytical solution of the problem with magnetic field in non-ideal medium. We have tried to fill the gap by providing approximate analytical solutions of the problem with magnetic and non-ideal effects. These findings are quite useful for the groups working in the field of Magneto Hydrodynamics (MHD) or Computational Fluid Dynamics (CFD). To generate confidence in our results, we have recovered all results of Sakurai's [4] published work as a particular case of our problem (i.e., $C_0 = 0$) in Tables 1 and 2. The effects of non-ideal medium and transverse magnetic field on flow variables are depicted in Figures 1–4.

Figure 1 consists of the profiles of velocity, density, pressure, and magnetic pressure for planar shocks ($m = 0$) with $\gamma = 1.4$, $C_0 = 0$, and $\alpha = 0.00, 0.015, 0.025, 0.050$. It is clearly seen from Figure 1 that the dimensionless profiles of flow variables decrease as value of parameter α increases except the pressure which increases with the small increase in the parameter α. This is expected physically as gas particles will collide more frequently with the increase of non-idealness α which results in the rise in pressure with the increment of α. In Figure 2, we observe that the pressure and density decrease as we increase the value of C_0, while the magnetic pressure increases for planar motion ($m = 0$) after a certain point. It is again an expected result as the charged particles will be transported away very quickly with the increase of C_0. It results in dropping in pressure after the blast. At the point of explosion, the process is not fully self-similar and therefore the initial dynamics is a bit off. It agrees with blast wave theory. It is well known that the phenomena are not self-similar at very near the center or axis of explosion. However, point explosions can be self-similar at considerable distance from the source (see Sakurai [3,4]). Figures 3 and 4 display the profiles of the flow variables for cylindrically symmetric flow ($m = 1$). From Figures 3a and 4a, we observed that as we increase the value of α and C_0, velocity increases. Figure 3b shows that as the value of parameter α increases, pressure also increases after a certain point. This describes the physics of the post-shock process very well as increase in α makes the collision of the gas particles more frequent. Figure 4b shows that an increment in the value of C_0 causes a decrease in pressure for the cylindrically symmetric flow. It is attributed to the fact that the ionized gas molecules are carried away by a strong magnetic field. It is one of the reasons that the instabilities at the shock front can be suppressed under the presence of a magnetic field. Figure 4c shows that the density of the medium decreases as C_0 increases. Figure 4d shows that as the value of C_0 increases, magnetic pressure also increases.

Author Contributions: All authors contributed equally in the paper.

Funding: This research was funded by University Grant Commission (UGC), India grant number "2121440656" and Ref. No; 21/12/2014(ii)EU-V.

Acknowledgments: Astha Chauhan is thankful to the "University Grant Commission", New Delhi for the financial support to project.

Conflicts of Interest: The authors declare no conflict of interest.

References

1. Sedov, L. Propagation of strong shock waves. *J. Appl. Math. Mech.* **1946**, *10*, 241–250.
2. Taylor, G. The formation of a blast wave by a very intense explosion. *Proc. R. Soc. A* **1950**, *201*, 159–174.
3. Sakurai, A. On the propagation and structure of the blast wave, I. *J. Phys. Soc. Japan* **1953**, *8*, 662–669. [CrossRef]
4. Sakurai, A. On the propagation and structure of a blast wave, II. *J. Phys. Soc. Japan* **1954**, *9*, 256–266. [CrossRef]

5. Murata, S. New exact solution of the blast wave problem in gas dynamics. *Chaos Solitons Fract.* **2006**, *28*, 327–330. [CrossRef]
6. Donato, A. Similarity analysis and non-linear wave propagation. *Int. J. Non-Linear Mech.* **1987**, *22*, 307–314. [CrossRef]
7. Book, D.L. The Sedov self-similar point blast solutions in non-uniform media. *Shock Waves* **1994**, *4*, 1–10. [CrossRef]
8. Lalicata, M.V.; Torrisi, M. Group analysis approach for a binary reacting mixture. *Int. J. Non-Linear Mech.* **1994**, *29*, 279–288. [CrossRef]
9. Anisimov, S.; Spiner, O. Motion of an almost ideal gas in the presence of a strong point explosion. *J. Appl. Math. Mech.* **1972**, *36*, 883–887. [CrossRef]
10. Lerche, I. Mathematical theory of cylindrical isothermal blast waves in a magnetic field. *Aust. J. Phys.* **1981**, *34*, 279–302. [CrossRef]
11. Lerche, I. Mathematical theory of one-dimensional isothermal blast waves in a magnetic field. *Aust. J. Phys.* **1979**, *32*, 491–502. [CrossRef]
12. Arora, R.; Yadav, S.; Siddiqui, M.J. Similarity method for the study of strong shock waves in magnetogasdynamics. *Bound. Value Probl.* **2014**, *2014*, 142. [CrossRef]
13. Siddiqui, M.J.; Arora, R.; Kumar, A. Shock waves propagation under the influence of magnetic field. *Chaos Solitons Fract.* **2017**, *97*, 66–74. [CrossRef]
14. Singh, L.; Husain, A.; Singh, M. On the evolution of weak discontinuities in radiative magnetogasdynamics. *Acta Astronaut.* **2011**, *68*, 16–21. [CrossRef]
15. Singh, L.; Husain, A.; Singh, M. An analytical study of strong non-planar shock waves in magnetogasdynamics. *Adv. Theor. Appl. Mech.* **2010**, *6*, 291–297.
16. Pandey, M.; Radha, R.; Sharma, V. Symmetry analysis and exact solutions of magnetogasdynamic equations. *Q. J. Mech. Appl. Math.* **2008**, *61*, 291–310. [CrossRef]
17. Menon, V.; Sharma, V. Characteristic wave fronts in magnetohydrodynamics. *J. Math. Anal. Appl.* **1981**, *81*, 189–203. [CrossRef]
18. Arora, R.; Siddiqui, M.J.; Singh, V. Similarity method for imploding strong shocks in a non-ideal relaxing gas. *Int. J. Non-Linear Mech.* **2013**, *57*, 1–9. [CrossRef]
19. Siddiqui, M.J.; Arora, R.; Singh, V. Propagation of non-linear waves in a non-ideal relaxing gas. *Int. J. Comput. Math.* **2018**, *95*, 2330–2342. [CrossRef]
20. Arora, R.; Siddiqui, M.J. Evolutionary behavior of weak shocks in a non-ideal gas. *J. Theor. Appl. Phys.* **2013**, *7*, 14. [CrossRef]
21. Ramu, A.; Dunna, N.; Satpathi, D.K. Numerical study of shock waves in non-ideal magnetogasdynamics (MHD). *J. Egypt. Math. Soc.* **2016**, *24*, 116–124. [CrossRef]
22. Whitham, G. Linear and nonlinear waves. *Phys. Today* **1975**, *28*, 55. [CrossRef]
23. Zhuravskaya, T.; Levin, V. The propagation of converging and diverging shock waves under intense heat exchange conditions. *J. Appl. Math. Mech.* **1996**, *60*, 745–752. [CrossRef]

© 2019 by the authors. Licensee MDPI, Basel, Switzerland. This article is an open access article distributed under the terms and conditions of the Creative Commons Attribution (CC BY) license (http://creativecommons.org/licenses/by/4.0/).

Article

Relationship between the Paradox of Enrichment and the Dynamics of Persistence and Extinction in Prey-Predator Systems

Jawdat Alebraheem

Mathematics Department, College of Science Al Zufli, Majmaah University, Majmaah 11952, Saudi Arabia; j.alebraheem@mu.edu.sa; Tel.: +966-53-5479321

Received: 3 September 2018; Accepted: 10 October 2018; Published: 22 October 2018

Abstract: The paradox of the enrichment phenomenon, considered one of the main counterintuitive observations in ecology, likely destabilizes predator–prey dynamics by increasing the nutrition of the prey. We use two systems to study the occurrence of the paradox of enrichment: The prey–predator system and the one prey, two predators system, with Holling type I and type II functional and numerical responses. We introduce a new approach that involves the connection between the occurrence of the enrichment paradox and persistence and extinction dynamics. We apply two main analytical techniques to study the persistence and extinction dynamics of two and three trophics, respectively. The linearity and nonlinearity of functional and numerical responses plays important roles in the occurrence of the paradox of enrichment. We derive the persistence and extinction conditions through the carrying capacity parameter, and perform some numerical simulations to demonstrate the effects of the paradox of enrichment when increasing carrying capacity.

Keywords: paradox of enrichment; prey–predator system; persistence of predators; extinction of predators

---

1. Introduction

Prey–predator interactions are important in applied mathematics and mathematical biology, receiving considerable attention from many researchers [1–8]. The Lotka-Volterra model is considered the basis for formulating prey–predator interaction models; it was proposed independently by Lotka and Volterra, so it is known as the Lotka-Volterra model. In the literature, predation and competition relationships are two main relationship types used for modeling any prey–predator system [9,10]. Mathematically, prey–predator interactions are described by nonlinear differential equations.

Counterintuitive observations have generally attracted more attention than observations that confirm intuition. These observations are called paradoxes that unexpectedly challenge normal intuition [11]. One of these observations, the paradox of enrichment, states that increasing the carrying capacity of prey in a stable prey–predator system leads to the destabilization of the system, which can be mathematically represented by limit cycles. Destabilization might lead to extinction, which is interpreted when the limit cycle is sufficiently large for one of the species or all species, so that the limit cycle is approximately close to zero. This phenomenon was discovered by Rosenzweig in 1971 [12].

Several experimental studies rejected the hypothesis that the enrichment phenomenon would destabilize community dynamics [13–16]. The studies that rejected enrichment paradox phenomenon explained that the paradox was actually caused by a difference between the mathematical construction and real prey–predator interactions. However, recent experimental studies showed the occurrence of the paradox of enrichment. Fussmann et al. [17] showed that enrichment led to the predator's extinction in their experiment on rotifer algae. Cottingham et al. [18] showed that in some lakes, anthropogenic eutrophication of ecosystems destabilized lakes. The process of lake eutrophication has been suggested to be an example of the paradox of enrichment [11]. Recently, Meyer et al. [19] predicted the occurrence of the paradox of enrichment for communities with multiple aboveground and belowground trophic levels and suggested that extinction and destabilization are more likely in fertilized agroecosystems than in natural communities.

One of the most important dynamics in prey–predator systems is stability, which is the first property usually studied in these systems. Some models show that the predator equilibrium density increases when the carrying capacity raises, but the prey equilibrium density would not increase as shown in the Rosenzweig–MacArthur model [20]. Notably, increasing the carrying capacity affects the prey and predator equilibrium densities. Losing stability transitions the dynamic behavior to cycle dynamics, which are relevant to persistence and extinction dynamics. The persistence and extinction of prey-predator systems have been studied by many researchers [21–41] due to their importance. Some methodologies have been used to find the conditions of persistence and extinction in two and three dimensions (trophics). Hutson and Vickers [23] determined the main criteria of a two prey, one predator model that depends on the Lyapunov function or average Lyapunov function. Freedman and Waltman [24] introduced a definition of persistence and determined the general criteria for three interacting populations. Freedman [21], in his book, summarized Kolmogorov conditions of prey–predator systems, which have been applied to derive the persistence and extinction conditions of two dimensions (trophics).

Persistence is defined analytically as follows: For a population $x(t)$, $if\ x(0) > 0$ and $\liminf_{t\to\infty} x(t) > 0$, $x(t)$ persists: geometrically, defined each trajectory of differential equations is defined as eventually bounding away from the coordinate planes [24]. Extinction is defined analytically as follows: if $x(0) > 0$ and $\liminf_{t\to\infty} x(t) = 0$, then $x(t)$ becomes extinct: geometrically, the trajectory of differential equations is defined as touching the coordinate planes.

Dubey and Upadhyay [30] studied persistence and extinction according to the Hutson and Vickers method. They explained that the conditions of persistence and extinction depend on the equilibrium levels of prey and predators and food conversion coefficients, capturing the rates and comparing them with the mortality rates of predators. Gakkhar et al. [31] studied persistence and extinction in their proposed model based on the Freedman and Waltman method. They proved that persistence is not possible for two predators competing for one prey species when any one of the boundary prey–predator planes has a stable equilibrium point. They presented numerical simulations of persistence in the case of periodic solution. They concluded that the principle of competitive exclusion holds in this case. Alebraheem and Abu Hassan [38–41] studied different scenarios of persistence and extinction in their modified model. However, the carrying capacity of the systems was widely excluded to study the dynamic behavior.

In this paper, we introduce a new approach that involves a mathematical connection between the occurrence of the enrichment paradox and the persistence and extinction dynamics. The question that we aimed to answer here is if enrichment of prey affects the persistence and extinction of predators. Therefore, we derived the persistence and extinction conditions and completed numerical simulations based on the carrying capacity that affects the occurrence of the paradox of enrichment. To study this idea, we used the same systems that were used by Alebraheem and Abu Hassan [38–42], but considered the carrying capacity. Two systems were examined: a prey–predator model that represents two dimensions (trophics), and a one prey, two predators system that describes three dimensions (trophics). Kolmogorov analysis and Freedman and Waltman methods were used to study the persistence and extinction dynamics.

The remainder of this paper is structured as follows. In Section 2, we introduce the mathematical systems of prey–predator used to study the relationship between the paradox of enrichment and the dynamics of persistence and extinction. In Section 3, we study the occurrence of the paradox of enrichment phenomenon. In Section 4, we study a theoretical approach to persistence and extinction. In Section 5, we present some numerical simulations. In Section 6, we draw our conclusions.

2. Mathematical Systems

In this paper, we introduce non-dimensional systems of two and three trophics. Holling type I and II functional and numerical responses are used to describe the predation of predators on prey and the effect of prey consumption on predators. Holling type I represents a linear function, whereas Holling type II represents a nonlinear function. The model can be formulated as:
The system of two trophics is as follows:

$$\frac{dx}{dt} = x\left(1 - \frac{x}{k}\right) - f(x)y,$$

$$\frac{dy}{dt} = -uy + R_1 y\left(1 - \frac{y}{k_y}\right),$$

(1)

with initial conditions

$$x(0) = x_0, \ (0) = y_0.$$

The system of three trophics is as follows:

$$\frac{dx}{dt} = x\left(1 - \frac{x}{k}\right) - f(x)y - g(x)z = xJ(x,y,z),$$

$$\frac{dy}{dt} = -uy + R_1 y\left(1 - \frac{y}{k_y}\right) - c_1 yz = yL_1(x,y,z),$$

(2)

$$\frac{dz}{dt} = -wz + R_2 z\left(1 - \frac{z}{k_z}\right) - c_2 yz = zL_2(x,y,z),$$

with initial conditions

$$x(0) = x_0, \ (0) = y_0, \ z(0) = z_0.$$

The different parameters in systems (1) and (2) are explained as follows. The intrinsic growth rate of prey is 1. In the case of Holling type I, $f(x) = \alpha x$ and $g(x) = \beta x$ are the functional responses to predators y and z, respectively, whereas in the case of Holling type II, $f(x) = \frac{\alpha x}{1+h_1 \alpha x}$ and $g(x) = \frac{\beta x}{1+h_2 \beta z}$ are the functional responses to predators y and z, respectively. For type I, the numerical responses are $R_1 = e_1 \alpha x$ and $R_2 = e_2 \beta x$ of the predators y and z, respectively. For Holling type II, $R_1 = \frac{e_1 \alpha x}{1+h_1 \alpha x}$ and $R_2 = \frac{e_2 \beta x}{1+h_2 \beta z}$. The parameters α and β measure the efficiency of the search and the capture of predators y and z, respectively. In the absence of prey x, the constants u and w are the death rates of predators y and z, respectively. h_1 and h_2 represent the handling and digestion rates of the predators, respectively, and e_1 and e_2 symbolize the efficiency of converting consumed prey into predator births. The carrying capacities $k_y = a_1 x$ and $k_z = a_2 x$ are proportional to the available amount of prey. In this paper, we assume $a_1 = a_2 = 1$ to simplify the mathematical analysis. c_1 and c_2 measure the interspecific competition between the predators. All the parameters and initial conditions of systems (1) and (2) are assumed to be positive values.

3. Occurrence of the Paradox of Enrichment

According to Jensen and Ginzburg [11], the paradox of enrichment is accepted intuition and must be considered a theory in ecology. In this section, we study the occurrence of the paradox of enrichment on systems (1) and (2) with Holling types I and II. To study this phenomenon, we discuss the stability of the coexistence equilibrium points $E = (x,y)$ and $E = (x,y,z)$ in two and three trophics, respectively.

3.1. Occurrence of the Paradox of Enrichment with Holling Type I

To check this phenomenon with Holling type I, we found the coexistence equilibrium points of systems (1) and (2) and present some theorems that prove the occurrence of the paradox of enrichment in systems (1) and (2) with Holling type I.

The coexistence equilibrium point of system (1) with Holling type I is $\hat{E} = (\hat{x}, \hat{y}) = \left(\frac{k(u+e)}{e+k e\alpha}, \frac{k e\alpha - u}{e\alpha + e\alpha^2 k}\right)$. It exists (positive equilibrium point) under the following condition:

$$k e\alpha > u \qquad (3)$$

$\bar{E} = (\bar{x}, \bar{y}, \bar{z})$ represents the coexistence of the equilibrium point of system (2) with Holling type I, which is obtained through the positive solution of the following algebraic system:

$$1 - \frac{x}{k} - \alpha y - \beta z = 0$$
$$-u + e_1 \alpha x - e_1 \alpha y - c_1 z = 0 \qquad (4)$$
$$-w + e_2 \beta x - e_2 \beta z - c_2 y = 0$$

Theorem 1. *The coexistence equilibrium point $\hat{E} = (\hat{x}, \hat{y}) = \left(\frac{k(u+e)}{e+k e\alpha}, \frac{k e\alpha - u}{e\alpha + e\alpha^2 k}\right)$ of system (1) is globally asymptotically stable within the positive quadrant of the $x - y$ plane.*

Proof. Let $G(x, y) = \frac{1}{xy}$. G is a Dulac function. It is continuously differentiable in the positive quadrant of the $x - y$ plane $A = \{\{(x, y) | x > 0, y > 0\}$ Hsu [43].

$$N_1(x, y) = x\left(1 - \frac{x}{k}\right) - \alpha xy,$$

$$N_2(x, y) = -uy + e_1 \alpha xy - e_1 \alpha y^2.$$

□.

Thus, $\Delta(GN_1, GN_2) = \frac{\partial(GN_1)}{\partial x} + \frac{\partial(GN_2)}{\partial y} = \frac{-1}{yk} - \frac{e_1 \alpha}{x}$.

It is observed that $\Delta(GN_1, GN_2)$ is not identically zero and does not change sign in the positive quadrant of the $x - y$ plane. Per the Bendixson–Dulac criterion, there is no periodic solution inside the positive quadrant of the $x - y$ plane. E_2 is globally asymptotically stable inside the positive quadrant of the $x - y$ plane.

Theorem 2. *The coexistence equilibrium point $\bar{E} = (\bar{x}, \bar{y}, \bar{z})$ of system (2) is globally asymptotically stable.*

Proof. The global stability of positive equilibrium point $\bar{E}$ is proved by using Lyapunov function.

$$V = B_1(x - \bar{x} - \ln(\tfrac{x}{\bar{x}})) + B_2(y - \bar{y} - \ln(\tfrac{y}{\bar{y}})) + B_3(z - \bar{z} - \ln(\tfrac{z}{\bar{z}})). \qquad (5)$$

□.

Differentiating V with respect to time along the solutions of the system (5)

$$\frac{dV}{dt} = B_1(x - \bar{x})\left[\left(1 - \frac{x}{k} - \alpha y - \beta z\right) - \left(1 - \frac{\bar{x}}{k} - \alpha \bar{y} - \beta \bar{z}\right)\right] + B_2(y - \bar{y})[(-u + e_1 \alpha x - e_1 \alpha y - c_1 z) \qquad (6)$$
$$-(-u + e_1 \alpha \bar{x} - e_1 \alpha \bar{y} - c_1 \bar{z})] + B_3(z - \bar{z})[(-w + e_2 \beta x - e_2 \beta z - c_2 y) - (-w + e_2 \beta \bar{x} - e_2 \beta \bar{z} - c_2 \bar{y})].$$

$$\frac{dV}{dt} = B_1(x - \bar{x})\left[\frac{-(x - \bar{x})}{k} - \alpha(y - \bar{y}) - \beta(z - \bar{z})\right] + B_2(y - \bar{y})[e_1 \alpha(x - \bar{x}) - e_1 \alpha(y - \bar{y}) - c_1(z - \bar{z})] \qquad (7)$$
$$+ B_3(z - \bar{z})[e_2 \beta(x - \bar{x}) - e_2 \beta(z - \bar{z}) - c_2(y - \bar{y})].$$

$$\frac{dV}{dt} = B_1 \left[\frac{-(x-\bar{x})^2}{k} - \alpha(x-\bar{x})(y-\bar{y}) - \beta(x-\bar{x})(z-\bar{z}) \right] + B_2 \left[\begin{array}{c} e_1\alpha(y-\bar{y})(x-\bar{x}) - e_1\alpha(y-\bar{y})^2 \\ -c_1(y-\bar{y})(z-\bar{z}) \end{array} \right] \quad (8)$$
$$+ B_3 \left[e_2\beta(z-\bar{z})(x-\bar{x}) - e_2\beta(z-\bar{z})^2 - c_2(z-\bar{z})(y-\bar{y}) \right].$$

By selecting $B_1 = 1$, $B_2 = \frac{1}{e_1}$, and $B_3 = \frac{1}{e_2}$, so

$$\frac{dV}{dt} = -\frac{1}{k}(x-\bar{x})^2 - \alpha(y-\bar{y})^2 - c_1(y-\bar{y})(z-\bar{z}) - \beta(z-\bar{z})^2 - c_2(z-\bar{z})(y-\bar{y}) \quad (9)$$

We conclude that $\frac{dV}{dt}$ is a negative definite without any conditions (i.e., no constraints on parameters).

In this section, the main result shows that the dynamic behaviors of systems (1) and (2) are always stable and there is no bifurcation under any conditions through Theorems 1 and 2. Therefore, the paradox of enrichment in systems (1) and (2) with Holling type I does not occur.

The system is stable, or population oscillations with small amplitude are likely to occur if the defense of prey is effective when compared with the predator's attacking [6].

3.2. Occurrence of the Paradox of Enrichment with Holling Type II

We found the coexistence equilibrium points of systems (1) and (2) and present some theorems that prove the occurrence of the paradox of enrichment in these systems with Holling type II.

The coexistence equilibrium point $\acute{E} = (\acute{x}, \acute{y})$ is obtained for system (1) with Holling type II through the positive root of the quadratic equation

$$\acute{x}^2 + \left(\frac{1}{h_1} - \frac{u}{e_1} - \frac{1}{k} + \frac{1}{h_1\alpha} \right) \acute{x} - \left(\frac{1}{h_1\alpha} + \frac{u}{e_1 h_1 \alpha} \right) = 0 \quad (10)$$

and

$$\acute{y} = \frac{1}{\alpha} \left(1 - \frac{\acute{x}}{k} \right) (1 + h_1 \alpha \acute{x}) \quad (11)$$

The coexistence equilibrium point $\bar{\bar{E}} = \left(\bar{\bar{x}}, \bar{\bar{y}}, \bar{\bar{z}} \right)$ of system (2) with Holling type II is obtained through the positive solution of the following algebraic system:

$$1 - \frac{x}{k} - \frac{\alpha y}{1 + h_1 \alpha x} - \frac{\beta z}{1 + h_2 \beta x} = 0$$

$$-u + \frac{e_1 \alpha x}{1 + h_1 \alpha x} - \frac{e_1 \alpha}{1 + h_1 \alpha x} y - c_1 z = 0 \quad (12)$$

$$-w + \frac{e_2 \beta x}{1 + h_2 \beta x} - \frac{e_2 \beta}{1 + h_2 \beta x} z - c_2 y = 0$$

To check this phenomenon with Holling type II, we present the following theorems:

Theorem 3. *The coexistence equilibrium point $\acute{E} = (\acute{x}, \acute{y})$ is asymptotically stable under the following condition:*

$$k < \frac{\acute{x}}{\frac{h_1 \alpha^2 \acute{y}}{(1+h_1 \alpha \acute{x})^2} - \frac{e_1 \alpha \acute{y}}{1+h_1 \alpha \acute{x}}} \quad (13)$$

Proof. The variational matrix of coexistence point $\acute{E}$ is as follows:

$$\acute{V} = \begin{pmatrix} \acute{x}(-\frac{1}{k} + \frac{h_1 \alpha^2 \acute{y}}{(1+h_1 \alpha \acute{x})^2}) & -\acute{x}(\frac{\alpha}{1+h_1 \alpha \acute{x}}) \\ \acute{y}(\frac{e_1 \alpha + h_1 e_1 \alpha^2 \acute{y}}{(1+h_1 \alpha \acute{x})^2}) & -\acute{y}(\frac{e_1 \alpha}{1+h_1 \alpha \acute{x}}) \end{pmatrix}$$

☐.

Through the variational matrix, the equilibrium point $\acute{E}$ is locally asymptotically stable, provided the following condition holds:

$$k < \frac{\acute{x}}{\frac{h_1 a^2 \acute{y}}{(1+h_1 a \acute{x})^2} - \frac{e_1 a \acute{y}}{1+h_1 a \acute{x}}}$$

Corollary 1. *If condition (13) is not satisfied, then the coexistence equilibrium point $\acute{E} = (\acute{x}, \acute{y})$ is unstable.*

Through Corollary 1, there is a destabilization of the coexistence equilibrium point $\acute{E}$ according to the carrying capacity parameter, so the paradox of enrichment in system (1) with Holling type II would occur. Therefore, the paradox of enrichment occurs in system (1) with Holling type II through some numerical simulations.

Theorem 4. *The coexistence equilibrium point $\bar{\bar{E}} = \left(\bar{\bar{x}}, \bar{\bar{y}}, \bar{\bar{z}} \right)$ of system (2) is obtained through the positive solution of system (12). It is locally asymptotically stable proven that conditions (15), (16), and (17) hold.*

The variational matrix of $\bar{\bar{E}}$ is as follows:

$$\bar{\bar{v}} = \begin{bmatrix} \bar{\bar{x}}(-\frac{1}{k} + \frac{h_1 a^2 \bar{\bar{y}}}{(1+h_1 a \bar{\bar{x}})^2} + \frac{h_2 \beta^2 \bar{\bar{z}}}{(1+h_2 \beta \bar{\bar{x}})^2}) & -\bar{\bar{x}}(\frac{\alpha}{1+h_1 a \bar{\bar{x}}}) & -\bar{\bar{x}}(\frac{\beta}{1+h_2 \beta \bar{\bar{x}}}) \\ \bar{\bar{y}}(\frac{e_1 a + h_1 e_1 a^2 \bar{\bar{y}}}{(1+h_1 a \bar{\bar{x}})^2}) & -\bar{\bar{y}}(\frac{e_1 a}{1+h_1 a \bar{\bar{x}}}) & -c_1 \bar{\bar{y}} \\ \bar{\bar{z}}(\frac{e_2 \beta + e_2 h_2 \beta^2 \bar{\bar{z}}}{(1+h_2 \beta \bar{\bar{x}})^2}) & -c_2 \bar{\bar{z}} & -\bar{\bar{z}}(\frac{e_2 \beta}{1+h_2 \beta \bar{\bar{x}}}) \end{bmatrix}$$

$$\bar{\bar{V}} = \begin{bmatrix} h_{11} & h_{12} & h_{13} \\ h_{21} & h_{22} & h_{23} \\ h_{31} & h_{32} & h_{33} \end{bmatrix}$$

where

$$h_{11} = \bar{\bar{x}}(-\frac{1}{k} + \frac{h_1 a^2 \bar{\bar{y}}}{(1+h_1 a \bar{\bar{x}})^2} + \frac{h_2 \beta^2 \bar{\bar{z}}}{(1+h_2 \beta \bar{\bar{x}})^2}), \ h_{12} = -\bar{\bar{x}}(\frac{\alpha}{1+h_1 a \bar{\bar{x}}}), h_{13} = -\bar{\bar{x}}(\frac{\beta}{1+h_2 \beta \bar{\bar{x}}})$$

$$h_{21} = \bar{\bar{y}}(\frac{e_1 a + h_1 e_1 a^2 \bar{\bar{y}}}{(1+h_1 a \bar{\bar{x}})^2}), \ h_{22} = -\bar{\bar{y}}(\frac{e_1 a}{1+h_1 a \bar{\bar{x}}}), \ h_{23} = -c_1 \bar{\bar{y}}$$

$$h_{31} = \bar{\bar{z}}(\frac{e_2 \beta + e_2 h_2 \beta^2 \bar{\bar{z}}}{(1+h_2 \beta \bar{\bar{x}})^2}), \ h_{32} = -c_2 \bar{\bar{z}}, \ h_{33} = -\bar{\bar{z}}(\frac{e_2 \beta}{1+h_2 \beta \bar{\bar{x}}}),$$

The characteristic equation of the variational matrix $\bar{\bar{V}}$ is as follows:

$$\lambda^3 + H_1 \lambda^2 + H_2 \lambda + H_3 = 0 \tag{14}$$

$$H_1 = -(h_{11} + h_{22} + h_{33})$$
$$H_2 = (h_{11}h_{22} + h_{23}h_{32} + h_{11}h_{33} + h_{22}h_{33} - h_{12}h_{21} - h_{13}h_{31})$$
$$H_3 = (h_{13}h_{31}h_{22} + h_{12}h_{21}h_{33} + h_{11}h_{23}h_{32} - h_{11}h_{22}h_{33} - h_{13}h_{21}h_{32} - h_{11}h_{22}h_{33})$$

According to Routh–Hurwitz criterion, $\bar{\bar{E}} = \left(\bar{\bar{x}}, \bar{\bar{y}}, \bar{\bar{z}}\right)$ is locally asymptotically stable if it holds the following conditions:

$$H_1 > 0 \tag{15}$$

$$H_3 > 0 \tag{16}$$

$$H_1 H_2 > H_3 \tag{17}$$

Theorem 5. *If one of the conditions (15)–(17) is not satisfied, then the coexistence equilibrium point $\bar{\bar{E}} = \left(\bar{\bar{x}}, \bar{\bar{y}}, \bar{\bar{z}}\right)$ is unstable.*

Proof. Through the variational matrix of coexistence point $\bar{\bar{E}} = \left(\bar{\bar{x}}, \bar{\bar{y}}, \bar{\bar{z}}\right)$, the stability is satisfied according to the Routh–Hurwitz criterion if all the conditions (15), (16), and (17) must be satisfied. However, it is observed from the first condition that

$$H_1 > 0 \text{ where } H_1 = -(h_{11} + h_{22} + h_{33})$$

□.

It is observed that $H_1 < 0$ when

$$\frac{h_1 \alpha^2 \bar{\bar{y}}}{\left(1 + h_1 \alpha \bar{\bar{x}}\right)^2} + \frac{h_2 \beta^2 \bar{\bar{z}}}{\left(1 + h_2 \beta \bar{\bar{x}}\right)^2} > \frac{\bar{\bar{x}}}{k} + \frac{e_1 \alpha \bar{\bar{y}}}{1 + h_1 \alpha \bar{\bar{x}}} + \frac{e_2 \beta \bar{\bar{z}}}{1 + h_2 \beta \bar{\bar{x}}} \tag{18}$$

Thus, the Routh–Hurwitz criterion is not satisfied, so the equilibrium point $\bar{\bar{E}}$ is unstable.

In this section, we concluded that the linearity and nonlinearity of functional and numerical responses plays important roles in the occurrence of the enrichment paradox. Some studies have shown that functional and numerical responses led to qualitative differences in dynamic behaviors of prey–predator systems [44–47].

Many biological factors control the shape of the functional and numerical responses as foraging theory and densities of prey, as shown in Nowak et al. [7]. Consequently, the shape of the functional and numerical responses affect the dynamic behaviors of prey–predator systems to be steady state, limit cycles, or complex dynamical behaviours.

4. Theoretical Approach to Persistence and Extinction

We studied persistence and extinction using different analytical techniques on systems (1) and (2). We introduce the conditions of persistence and extinction depending on the carrying capacity parameter. Therefore, we have four cases as follows:

In two dimensions, we use the Kolmogorov analysis to find the conditions of persistence and extinction.

For Holling type I, the persistence condition is as follows:

$$0 < \frac{u}{e\alpha} < k \tag{19}$$

However, if condition (19) is not satisfied to become as follows:

$$\frac{u}{e\alpha} \geq k \tag{20}$$

Then, the predator tends to be extinct.

For Holling type II, the persistence condition is as follows:

$$0 < \frac{u}{e\alpha - uh\alpha} < k \tag{21}$$

However, if

$$\frac{u}{e\alpha - uh\alpha} \geq k \tag{22}$$

Then, the predator tends to be extinct.

In three dimensions, some theorems must be proven for finding the persistence and extinction conditions of system (2) with Holling type I and those of system (2) with Holling type II in the case of nonperiodic solutions. However, the persistence conditions of the case of periodic solutions cannot be derived theoretically according to Freedman and Waltman [24], so we used the numerical simulations to show the probability of persistence and extinction cases.

Theorem 6. *The equilibrium point $\hat{E} = (\hat{x}, \hat{y}, 0) = \left(\frac{k(u+e_1)}{e_1+ke_1\alpha}, \frac{ke_1\alpha-u}{e_1\alpha+e_1\alpha^2 k}, 0\right)$ is unstable in the z-direction (i.e., orthogonal to the $x - y$ plane), if the following condition is satisfied:*

$$w + c_2\hat{y} < e_2\beta\hat{x} \tag{23}$$

Proof. The variational matrix of equilibrium point $\hat{E} = (\hat{x}, \hat{y}, 0)$ is computed as follows:

$$\hat{V} = \begin{pmatrix} -\frac{\hat{x}}{k} & -\alpha\hat{x} & -\beta\hat{x} \\ e_1\alpha\hat{y} & -e_1\alpha\hat{y} & -c_1\hat{y} \\ 0 & 0 & -w + e_2\beta\hat{x} - c_2\hat{y} \end{pmatrix}$$

□.

From $\hat{V}$ and by using the Routh–Hurwitz criterion, equilibrium point $\hat{E}$ is locally asymptotically stable, provided the following conditions hold:

$$w + c_2\hat{y} > e_2\beta\hat{x} \tag{22}$$

The equilibrium point $\hat{E}$ is stable in the $x - y$ plane if condition (24) is satisfied, so $\hat{E}$ is unstable in the z-direction (i.e., orthogonal to the $x - y$ plane) if condition (24) is not satisfied, which produces condition (23).

Theorem 7. *The equilibrium point $\tilde{E} = (\tilde{x}, 0, \tilde{z}) = \left(\frac{k(w+e_2)}{e_2+ke_2\beta}, 0, \frac{ke_2\beta-w}{e_2\beta+e_2\beta^2 k}\right)$ is unstable in the y-direction (i.e., orthogonal to the $x - z$ plane), if the following condition is satisfied:*

$$u + c_1\tilde{z} < e_1\alpha\tilde{x} \tag{23}$$

Proof. Following the same process, we prove this theorem along with Theorem 6, so the variational matrix of equilibrium point $\tilde{E}$ is as follows:

$$\dot{V} = \begin{pmatrix} -\frac{\tilde{x}}{k} & -\alpha\tilde{x} & -\beta\tilde{x} \\ 0 & -u + e_1\alpha\tilde{x} - c_1\tilde{z} & 0 \\ e_2\beta\tilde{z} & -c_2\tilde{z} & -e_2\beta\tilde{z} \end{pmatrix}$$

□.

From $\tilde{V}$ and using the Routh–Hurwitz criterion, equilibrium point $\tilde{E}$ is locally asymptotically stable, provided the following condition holds:

$$u + c_1 \tilde{z} > e_1 \alpha \tilde{x} \tag{26}$$

The equilibrium point $\tilde{E}$ is stable in the $x - z$ plane if condition (26) is satisfied, so $\dot{E}$ is unstable in the y-direction (i.e., orthogonal to the $x - z$ plane) if condition (26) is not satisfied, which produces condition (25).

Theorem 8. *System (2) with Holling type I is persistent if the following conditions hold:*

$$k \geq \frac{ue_2\beta - c_1 w}{e_1 \alpha \beta w - ue_2 \beta^2 + e_1 e_2 \alpha \beta - c_1 e_2 \beta} \tag{27}$$

$$k \geq \frac{we_1 \alpha - c_2 u}{e_2 \alpha \beta u - we_1 \beta^2 + e_1 e_2 \alpha \beta - c_2 e_1 \alpha} \tag{28}$$

Proof. As functions J, F_i; $i = 1, 2$ of system (2) are continuous in the positive volume $R_+^3 = \{(x, y, z) : x \geq 0, y \geq 0, z \geq 0\}$, the system is bounded with positive initial conditions because the prey is bounded, where $k > 0$ and the growth of predators depends on the prey. The conditions $L_1(\tilde{x}, 0, \tilde{z}) > 0$ and $L_2(\hat{x}, \hat{y}, 0) > 0$ are exactly needed to make the equilibrium points unstable in the orthogonal of the other coordinate planes (Theorems 6 and 7). System (2) has a nonperiodic solution only (i.e., no limit cycles) through Theorem 2. □

To complete the proof, the following hypotheses are satisfied with Freedman's and Waltman's theorem.

Hypothesis 1 (H1). $\frac{\partial J}{\partial y} = -\alpha < 0; \frac{\partial J}{\partial z} = -\beta < 0, \frac{\partial L_1}{\partial x} = e_1 \alpha > 0; \frac{\partial L_2}{\partial x} = e_2 \beta > 0$,
$L_1(0, y, z) = -u - e_1 \alpha y - c_1 z < 0; L_2(0, y, z) = -w - e_2 \beta z - c_2 y < 0$,
$\frac{\partial L_1}{\partial y} = -e_1 \alpha \leq 0; \frac{\partial L_1}{\partial z} = -c_1 \leq 0; \frac{\partial L_2}{\partial y} = -c_2 \leq 0; \frac{\partial L_2}{\partial z} = -e_2 \beta \leq 0$

Hypothesis 2 (H2). *If the predator is absent, then the prey species x growths to carrying capacity, i.e.,* $J(0, 0, 0) = 1 > 0, \frac{\partial J}{\partial x}(x, y, z) = \frac{-1}{k} \leq 0, \exists k > 0 \ni J(k, 0, 0) = 0, \ni J(k, 0, 0) = 0.$

Hypothesis 3 (H3). *There are no equilibrium points on the y or z coordinate axes and no equilibrium point in the y–z plane.*

Hypothesis 4 (H4). *The predator y and the predator z can survive on the prey, there exist points $\acute{E} = (\acute{x}, \acute{y}, 0)$ and $\widehat{E} = (\widehat{x}, 0, \widehat{z})$, such that $J(\acute{x}, \acute{y}, 0) = L_1(\acute{x}, \acute{y}, 0) = 0$ and $J(\widehat{x}, 0, \widehat{z}) = L_2(\ddot{x}, 0, \ddot{z}) = 0, \acute{x}, \acute{y}, \widehat{x}, \widehat{z} > 0$ and $\acute{x} < k, \widehat{x} < k$.*

Corollary 2. *The first predator y is extinct of system (2) with Holling type I if the following condition is satisfied:*

$$k < \frac{ue_2\beta - c_1 w}{e_1 \alpha \beta w - ue_2 \beta^2 + e_1 e_2 \alpha \beta - c_1 e_2 \beta} \tag{29}$$

Corollary 3. *The second predator z is extinct of system (2) with Holling type I if the following condition is satisfied:*

$$k < \frac{we_1 \alpha - c_2 u}{e_2 \alpha \beta u - we_1 \beta^2 + e_1 e_2 \alpha \beta - c_2 e_1 \alpha} \tag{30}$$

As such, we used the same technique to find the persistence and extinction of system (2) with Holling type II.

The equilibrium point $\acute{E} = (\acute{x}, \acute{y}, 0)$ of system (2) with Holling type II is obtained through the positive root of the following quadratic equation:

$$\acute{x}^2 + \left(\frac{1}{h_1} - \frac{u}{e_1} - \frac{1}{k} + \frac{1}{h_1 \alpha}\right)\acute{x} - \left(\frac{1}{h_1 \alpha} + \frac{u}{e_1 h_1 \alpha}\right) = 0 \tag{31}$$

and

$$\acute{y} = \frac{1}{\alpha}\left(1 - \frac{\acute{x}}{k}\right)(1 + h_1 \alpha \acute{x}) \tag{32}$$

Theorem 9. *The equilibrium point $\acute{E} = (\acute{x}, \acute{y}, 0)$ is unstable in the z-direction (i.e., orthogonal to the x − y plane) if the following condition is satisfied:*

$$w + c_2 \acute{y} < \frac{e_2 \beta \acute{x}}{1 + h_2 \beta \acute{x}} \tag{33}$$

Proof. The variational matrix of equilibrium point $\acute{E} = (\acute{x}, \acute{y}, 0)$ is computed as follows:

$$\acute{V} = \begin{pmatrix} \acute{x}\left(-\frac{1}{k} + \frac{h_1 \alpha^2 \acute{y}}{(1+h_1 \alpha \acute{x})^2}\right) & -\frac{\acute{x}\alpha}{1+h_1 \alpha \acute{x}} & -\frac{\acute{x}\beta}{1+h_2 \beta \acute{x}} \\ \frac{e_1 \alpha \acute{y}}{(1+h_1 \alpha \acute{x})^2} & -\frac{e_1 \alpha \acute{y}}{1+h_1 \alpha \acute{x}} & -c_1 \acute{y} \\ 0 & 0 & -w + \frac{e_2 \beta \acute{x}}{1+h_2 \beta \acute{x}} - c_2 \acute{y} \end{pmatrix}$$

□.

From $\acute{V}$ and using the Routh–Hurwitz criterion, equilibrium point $\acute{E}$ is locally asymptotically stable, provided the following conditions hold:

$$w + c_2 \acute{y} > \frac{e_2 \beta \acute{x}}{1 + h_2 \beta \acute{x}} \tag{34}$$

The equilibrium point $\acute{E}$ is stable in the $x - y$ plane if condition (34) is satisfied, so $\acute{E}$ is unstable in the z-direction (i.e., orthogonal to the $x - y$ plane) if condition (34) is not satisfied, which produces condition (33).

The equilibrium point $\widehat{E} = (\widehat{x}, 0, \widehat{z})$ of system (2) with Holling type II is obtained through the positive root of the quadratic equation as follows:

$$\widehat{x}^2 + \left(\frac{1}{h_2} - \frac{w}{e_2} - \frac{1}{k} + \frac{1}{h_2 \beta}\right)\widehat{x} - \left(\frac{1}{h_2 \beta} + \frac{w}{e_2 h_2 \beta}\right) = 0 \tag{35}$$

and

$$\widehat{z} = \frac{1}{\beta}\left(1 - \frac{\widehat{x}}{k}\right)(1 + h_2 \beta \widehat{x}) \tag{36}$$

Theorem 10. *The equilibrium point $\widehat{E} = (\widehat{x}, 0, \widehat{z})$ is unstable in the y-direction (i.e., orthogonal to the x − z plane) if the following condition is satisfied:*

$$u + c_1 \widehat{z} < \frac{e_1 \alpha \widehat{x}}{1 + h_1 \alpha \widehat{x}} \tag{37}$$

Proof. Following the same process, we proved this theorem with Theorem 9, so the variational matrix of equilibrium point $\widehat{E}$ is as follows:

$$\ddot{V} = \begin{pmatrix} \widehat{x}\left(-\frac{1}{k} + \frac{h_2\beta^2\, \widehat{z}}{\left(1+h_2\beta\, \widehat{x}\right)^2}\right) & -\frac{\widehat{x}\,\alpha}{1+h_1\alpha\, \widehat{x}} & -\frac{\widehat{x}\,\beta}{1+h_2\beta\, \widehat{x}} \\ 0 & -u + \frac{e_1\alpha\, \widehat{x}}{1+h_1\alpha\, \widehat{x}} - c_1\, \widehat{z} & 0 \\ \frac{e_2\beta\, \widehat{z}}{\left(1+h_2\beta\, \widehat{x}\right)^2} & -c_2\, \widehat{z} & -\frac{e_2\beta\, \widehat{z}}{1+h_1\alpha\, \widehat{x}} \end{pmatrix}$$

□.

From $\ddot{V}$ and using the Routh–Hurwitz criterion, equilibrium point $\ddot{E}$ is locally asymptotically stable, provided the following condition holds:

$$u + c_1\, \widehat{z} > \frac{e_1\alpha\, \widehat{x}}{1+h_1\alpha\, \widehat{x}} \tag{38}$$

The equilibrium point $\widehat{E}$ is stable in the $x - z$ plane if condition (38) is satisfied, so $\ddot{E}$ is unstable in the y-direction (i.e., orthogonal to the $x - z$ plane) if condition (38) is not satisfied, which produces condition (37).

We introduce the persistence conditions of system (2) with Holling type II in the nonperiodic dynamic system through the following theorem:

Theorem 11. *System (2) with Holling type II is persistent if the following conditions hold:*

$$-u + \frac{e_1 \alpha \ddot{x}}{1+h_1\alpha\ddot{x}} - c_1\, \widehat{z} \geq 0 \tag{39}$$

$$-w + \frac{e_2 \beta \dot{x}}{1+h_2\beta\dot{x}} - c_2\dot{y} \geq 0 \tag{40}$$

Proof. As functions J, F_i; $i = 1, 2$ of system (2) are continuous in the positive volume $R^3_+ = \{(x, y, z) : x \geq 0, y \geq 0, z \geq 0\}$, the system is bounded with positive initial conditions because the prey is bounded, where $k > 0$ and the growth of predators depends on the prey. The conditions $L_1\left(\widehat{x}, 0, \widehat{z}\right) > 0$ and $L_2(\hat{x}, \hat{y}, 0) > 0$ are exactly needed to make the equilibrium points become unstable in the orthogonal of the other coordinate planes (Theorems 9 and 10). System (2) has a nonperiodic solution (i.e., no limit cycles) through Theorem 4. □

To complete the proof, the following hypotheses are satisfied with Freedman's and Waltman's theorem.

We use $y_1 \equiv y$ and $y_2 \equiv z$ to simplify the notations.

Hypothesis 5 (H5). $\frac{\partial J}{\partial y_i} < 0$, $\frac{\partial L_i}{\partial x} > 0$, $L_i(0, y, z) < 0$, $\frac{\partial L_i}{\partial y_j} \leq 0\, i, j = 1, 2$.

Hypothesis 6 (H6). *in the absence of a predator, the prey species x growths to carrying capacity, i.e.,* $J(0, 0, 0) = 1 > 0$, $\frac{\partial J}{\partial x}(x, y, z) = -\frac{1}{k} \leq 0, \exists k > 0 \ni J(k, 0, 0) = 0, \ni J(k, 0, 0) = 0$.

Hypothesis 7 (H7). *There are no equilibrium points on the y or z coordinate axes and no equilibrium point in the y–z plane.*

Hypothesis 8 (H8). *The predator y and the predator z can survive on the prey, there exist points $\acute{E} = (\acute{x}, \acute{y}, 0)$ and $\widehat{E} = (\widehat{x}, 0, \widehat{z})$, such that $J(\acute{x}, \acute{y}, 0) = L_1(\acute{x}, \acute{y}, 0) = 0$ and $J(\widehat{x}, 0, \widehat{z}) = L_2(\ddot{x}, 0, \ddot{z}) = 0$, $\acute{x}, \acute{y}, \widehat{x}, \widehat{z} > 0$ and $\acute{x} < k$, $\widehat{x} < k$.*

Corollary 4. *The first predator y is extinct of system (2) with Holling type II if the following condition is satisfied:*

$$-u + \frac{e_1 \alpha \widehat{x}}{1 + h_1 \alpha \widehat{x}} - c_1 \widehat{z} < 0 \tag{41}$$

Corollary 5. *The second predator z is extinct of system (2) with Holling type II if the following condition is satisfied:*

$$-w + \frac{e_2 \beta \acute{x}}{1 + h_2 \beta \acute{x}} - c_2 \acute{y} < 0 \tag{42}$$

However, the persistence and extinction conditions of system (2) with Holling type II are not written in terms of the carrying capacity parameter (k), because writing the persistence conditions in this term is difficult where $\acute{x}$ and $\ddot{x}$ are obtained through the positive solutions of quadratic Equations (31)–(36), which involve the carrying capacity parameter (k).

In this section, we obtained the persistence and extinction conditions of systems (1) and (2) based on carrying capacity. In two dimensions, we applied Kolmogorov analysis to find persistence and extinction conditions (19)–(22) of system (1) and (2) with Holling type I and II, respectively. In three dimensions, we applied the Freedman and Waltman method [24] to obtain persistence and extinction conditions (27)–(30) of system (2) with Holling type I, and persistence and extinction conditions (39)–(42) of system (2) with Holling type II in the case of nonperiodic solutions.

Some experimental studies found that the carrying capacity has an important influence on persistence and extinction in experimental populations, as shown by Griffen and Drake [8].

5. Numerical Simulation

In this section, we present some numerical simulations to show the occurrence of the paradox of enrichment of systems (1) and (2) with Holling type II when increasing the carrying capacity of prey. We use time series and phase space graphs to present the dynamic behavior, and present bifurcation diagrams to explain a map of the dynamic behaviors of systems (1) and (2). The values of the parameters for both systems (1) and (2) were selected to satisfy Theorems 3 and 4, in which the dynamic behavior is stable, but different values of carrying capacity are used.

The values of system (1) are as follows:

$$\alpha = 16.0,\ e_1 = 0.7,\ h_1 = 0.5,\ u = 0.65,\ x(0) = 0.5,\ y(0) = 0.2 \tag{43}$$

The values of system (2) are as follows:

$$\alpha = 10.0,\ \beta = 7.0,\ e_1 = 1.00,\ e_2 = 0.4,\ h_1 = 1.5,\ h_2 = 1.7,\ c_1 = 0.03,\ c_2 = 0.02,$$
$$u = 0.05,\ w = 0.1,\ x(0) = 0.5,\ y(0) = 0.2,\ z(0) = 0.2 \tag{44}$$

When taking the value of $k = 1$ of system (1), the dynamic behavior in the first case is stable, as shown in Figure 1. However, when increasing the carrying capacity to $k = 4$ in the second case, the dynamic behavior oscillates for a period of time and then ends, finally stabilizing, as shown in Figure 2. In the third case, when $k = 7$, the dynamic behavior oscillates to become a limit cycle, as shown in Figure 3. Consequently, the probability of extinction in the third case would be higher than in the first and second cases. Figure 4 shows the changes of the dynamic behavior of system (1) with Holling type II, from stable to periodic cases. The points in Figure 4 appear because oscillation exists in the dynamic behavior.

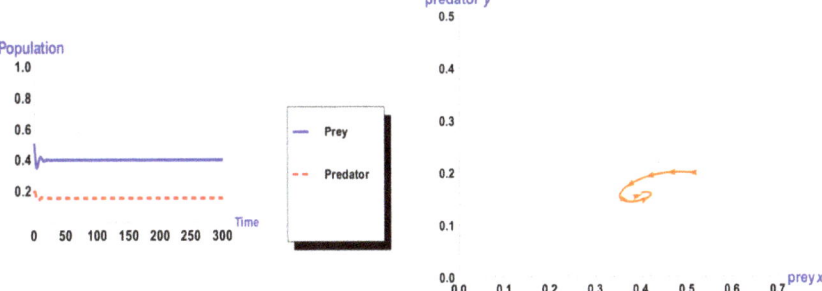

Figure 1. Dynamic behavior of system (1) when $k = 1$: (**a**) time series of two trophics x and y; (**b**) phase space of two trophics.

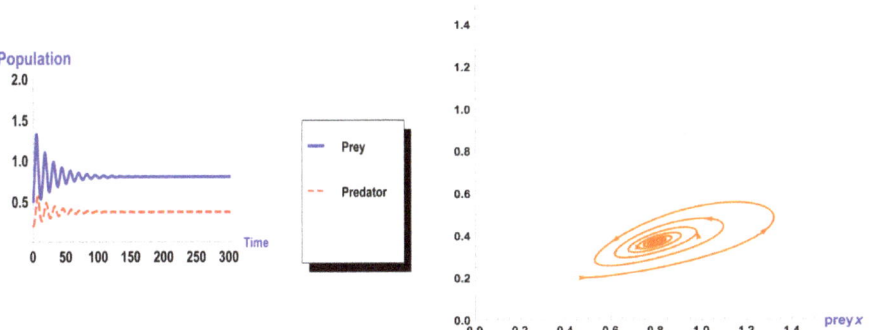

Figure 2. Dynamic behavior of system (1) when $k = 4$: (**a**) time series of two trophics x and y; (**b**) phase space of two trophics.

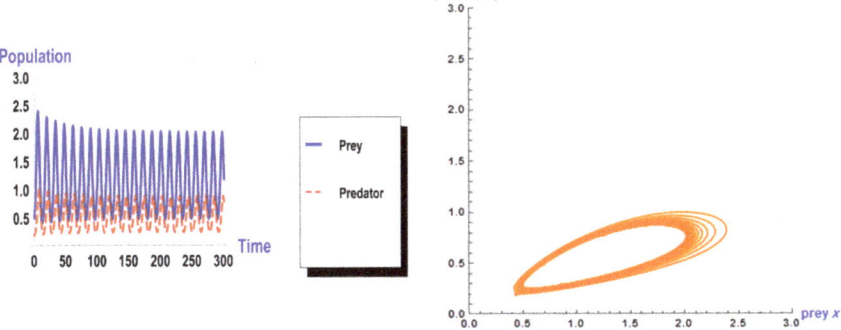

Figure 3. Dynamic behavior of system (1) when $k = 7$: (**a**) time series of two trophics x and y; (**b**) phase space of two trophics.

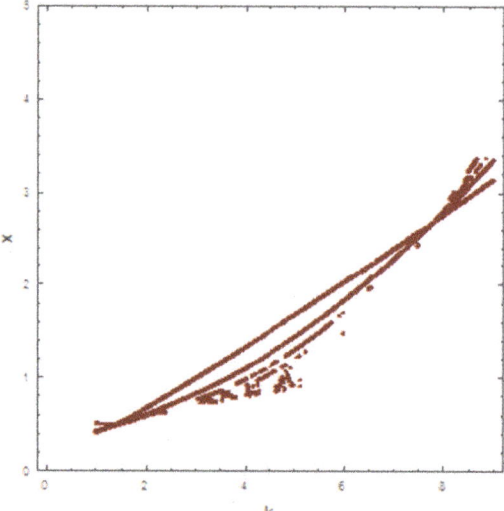

Figure 4. Bifurcation diagram for system (1) with Holling type II, using carrying capacity (k) as the bifurcation parameter.

Following the same process, we used different values of k from system (2). As shown in Figure 5, the dynamic behavior is stable when k =1 in the first case. However, when the carrying capacity is increased to k = 2 in the second case, the dynamic behavior oscillates for a period of time and then ends, finally stabilizing, as shown in Figure 6. Whereas in the third case, when k = 3, the dynamic behavior oscillates to create a limit cycle, as shown in Figure 7. Therefore, the probability of extinction in the third case would be greater than in the first and second cases. Figure 8 shows the changes in the dynamic behavior of system (2) with Holling type II, from stable to periodic, quasi-periodic, or chaos cases. The points in Figure 8 appear because oscillation occurs in the dynamic behavior. The numerical simulations show the occurrence of the paradox of enrichment in systems (1) and (2) with Holling type II.

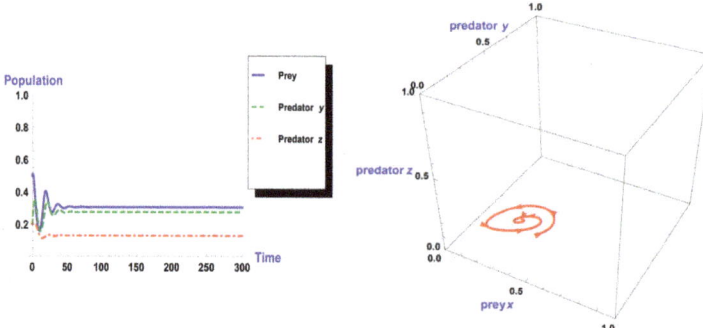

Figure 5. Dynamic behavior of system (2) when $k = 1$: (**a**) time series of three trophics x, y and z; (**b**) phase space of three trophics.

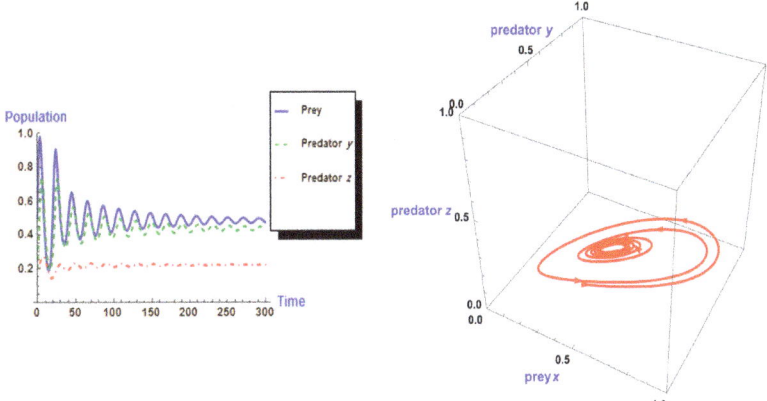

Figure 6. Dynamic behavior of system (2) when $k = 2$: (**a**) time series of three trophics x, y and z; (**b**) phase space of three trophics.

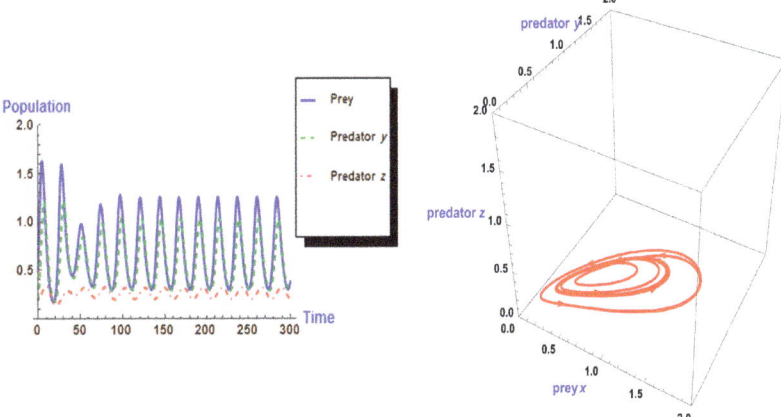

Figure 7. Dynamic behavior of system (2) when $k = 3$: (**a**) time series of three trophics x, y and z; (**b**) phase space of three trophics.

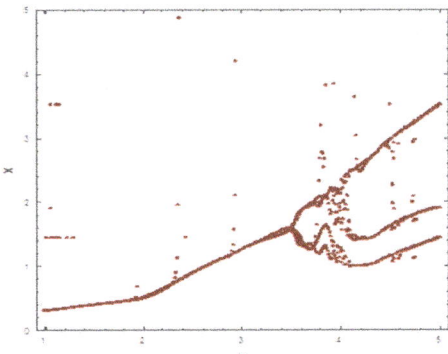

Figure 8. Bifurcation diagram for system (2) with Holling type II, using carrying capacity (k) as the bifurcation parameter.

6. Conclusions

We studied the occurrence of the paradox of enrichment in prey–predator models with Holling types I and II functional and numerical responses. We proved through Theorems 1 and 2 that the paradox of enrichment does not occur with Holling type I in two or three dimensions. However, the paradox of enrichment occurs with Holling type II in two and three dimensions, respectively, as shown through Corollary 1, Theorem 5, and the numerical simulations. The numerical simulations explain the occurrence of the paradox of enrichment in systems (1) and (2) with Holling type II when the carrying capacity of prey increases and a map of changes of the dynamic behaviors is given for stable to periodic, quasi-periodic, or chaos cases. We conclude that the linearity and nonlinearity of functional and numerical responses plays important roles in the occurrence of the enrichment paradox. We introduce a new approach connecting the enrichment paradox phenomenon and persistence and extinction dynamics by deriving the persistence and extinction conditions based on the carrying capacity parameter (k). We used different analytical techniques to derive the persistence and extinction conditions. We introduce several theorems and corollaries to present our results. We introduce some biological explanations to support our results.

Acknowledgments: This work is funded by the Basic Science Research Unit, Scientific Research Deanship at Majmaah University under the research project No. 31/37. The author is extremely grateful to Majmaah University, Deanship of Scientific Research and Basic Science Research Unit, Majmaah University.

Conflicts of Interest: The author declares no conflict of interest. The funders had no role in the design of the study; in the collection, analyses, or interpretation of data; in the writing of the manuscript, and in the decision to publish the results.

References

1. Liu, H.; Cheng, H. Dynamic analysis of a prey–predator model with state-dependent control strategy and square root response function. *Adv. Differ. Equ.* **2018**, *2018*, 63. [CrossRef]
2. Gurubilli, K.K.; Srinivasu, P.D.N.; Banerjee, M. Global dynamics of a prey-predator model with Allee effect and additional food for the predators. *Int. J. Dyn. Control* **2017**, *5*, 903–916. [CrossRef]
3. Keong, A.T.; Safuan, H.M.; Jacob, K. Dynamical behaviours of prey-predator fishery model with harvesting affected by toxic substances. *Matematika* **2018**, *34*, 143–151. [CrossRef]
4. Heurich, M.; Zeis, K.; Küchenhoff, H.; Müller, J.; Belotti, E.; Bufka, L.; Woelfing, B. Selective predation of a stalking predator on ungulate prey. *PLoS ONE* **2016**, *11*, e0158449. [CrossRef] [PubMed]
5. Gervasi, V.; Nilsen, E.B.; Sand, H.; Panzacchi, M.; Rauset, G.R.; Pedersen, H.C.; Kindberg, J.; Wabakken, P.; Zimmermann, B.; Odden, J.; et al. Predicting the potential demographic impact of predators on their prey: A comparative analysis of two carnivore–ungulate systems in Scandinavia. *J. Anim. Ecol.* **2012**, *81*, 443–454. [CrossRef] [PubMed]
6. Mougi, A.; Iwasa, Y. Evolution towards oscillation or stability in a predator-prey system. *Proc. Biol. Sci.* **2010**, *277*, 3163–3171. [CrossRef] [PubMed]
7. Nowak, E.M.; Theimer, T.C.; Schuett, G.W. Functional and numerical responses of predators: Where do vipers fit in the traditional paradigms? *Biol. Rev. Camb. Philos. Soc.* **2008**, *83*, 601–620. [CrossRef] [PubMed]
8. Griffen, B.D.; Drake, J.M. Effects of habitat quality and size on extinction in experimental populations. *Proc. R. Soc. B Biol. Sci.* **2008**, *275*, 2251–2256. [CrossRef]
9. Haberman, R. *Mathematical Models Mechanical Vibrations, Population Dynamics, and Traffic Flow*; Society for Industrial and Applied Mathematics: Philadelphia, PA, USA, 1998.
10. Murray, J.D. *Mathematical Biology*; Springer: New York, NY, USA, 2002; Volume 2.
11. Jensen, C.X.J.; Ginzburg, L.R. Paradoxes or theoretical failures? The jury is still out. *Ecol. Model.* **2005**, *188*, 3–14. [CrossRef]
12. Rosenzweig, M.L. Paradox of enrichment—Destabilization of exploitation ecosystems in ecological time. *Science* **1971**, *171*, 385–387. [CrossRef] [PubMed]
13. Walters, C.J.; Krause, E.; Neill, W.E.; Northcote, T.G. Equilibrium-models for seasonal dynamics of plankton biomass in 4 oligotrophic lakes. *Can. J. Fish. Aquat. Sci.* **1987**, *44*, 1002–1017. [CrossRef]

14. McCauley, E.; Murdoch, W.W. Predator prey dynamics in environments rich and poor in nutrients. *Nature* **1990**, *343*, 455–457. [CrossRef]
15. Persson, L.; Johansson, L.; Andersson, G.; Diehl, S.; Hamrin, S.F. Density dependent interactions in lake ecosystems—Whole lake perturbation experiments. *Oikos* **1993**, *66*, 193–208. [CrossRef]
16. Mazumder, A. Patterns of algal biomass in dominant odd-link vs. even-link lake ecosystems. *Ecology* **1994**, *75*, 1141–1149. [CrossRef]
17. Fussmann, G.F.; Ellner, S.P.; Shertzer, K.W.; Hairston, N.G., Jr. Crossing the hopf bifurcation in a live predator–prey system. *Science* **2000**, *290*, 1358–1360. [CrossRef] [PubMed]
18. Cottingham, K.L.; Rusak, J.A.; Leavitt, P.R. Increased ecosystem variability and reduced predictability following fertilisation: Evidence from palaeolimnology. *Ecol. Lett.* **2000**, *3*, 340–348. [CrossRef]
19. Meyer, K.M.; Vos, M.; Mooij, W.M.; Hol, W.H.G.; Termorshuizen, A.J.; van der Putten, W.H. Testing the Paradox of Enrichment along a Land Use Gradient in a Multitrophic Aboveground and Belowground Community. *PLoS ONE* **2012**, *7*, e49034. [CrossRef] [PubMed]
20. Oksanen, L.; Fretwell, S.D.; Arruda, J.; Niemela, P. Exploitation ecosystems in gradients of primary productivity. *Am. Nat.* **1981**, *118*, 240–261. [CrossRef]
21. Freedman, I. *Deterministic Mathematical Models in Population Ecology*; Marcel Dekker, Inc.: New York, NY, USA, 1980.
22. Smith, H.L. Competitive coexistence in an oscillating chemostat. *SIAM J. Appl. Math.* **1981**, *40*, 498–522. [CrossRef]
23. Hutson, V.; Vickers, G.A. Criterion for permanent coexistence of species, with an application to a two-prey one-predator system. *Math. Biosci.* **1983**, *63*, 253–269. [CrossRef]
24. Freedman, H.; Waltman, P. Persistence in models of three interacting predator-prey populations. *Math. Biosci.* **1984**, *68*, 213–231. [CrossRef]
25. Waltman, P. Coexistence in chemostat-like models. *Rocky Mt. J. Math.* **1990**, *20*, 777–807. [CrossRef]
26. Ruan, S.; Freedman, H.I. Persistence in three-species food chain models with group defence. *Math. Biosci.* **1991**, *107*, 111–125. [CrossRef]
27. Kuang, Y.; Beretta, E. Global qualitative analysis of a ratio-dependent predator-prey system. *J. Math. Biol.* **1998**, *36*, 389–406. [CrossRef]
28. Kuang, Y. Basic properties of mathematical population models. *J. Biomath.* **2002**, *17*, 129–142.
29. Hsu, S.-B.; Hwang, T.-W.; Kuang, Y. A ratio-dependent food chain model and its applications to biological control. *Math. Biosci.* **2003**, *181*, 55–83. [CrossRef]
30. Dubey, B.; Upadhyay, R. Persistence and extinction of one-prey and two-predator system. *Nonlinear Anal.* **2004**, *9*, 307–329.
31. Gakkhar, S.; Singh, B.; Naji, R.K. Dynamical behavior of two predators competing over a single prey. *Biosystems* **2007**, *90*, 808–817. [CrossRef] [PubMed]
32. Naji, R.K.; Balasim, A.T. Dynamical behavior of a three species food chain model with beddington-deangelis functional response. *Chaos Solitons Fractals* **2007**, *32*, 1853–1866. [CrossRef]
33. Upadhyay, R.K.; Naji, R.K. Dynamics of a three species food chain model with crowley-martin type functional response. *Chaos Solitons Fractals* **2009**, *42*, 1337–1346. [CrossRef]
34. Huo, H.F.; Ma, Z.P.; Liu, C.Y. Persistence and stability for a generalized leslie-gower model with stage structure and dispersal. *Abstr. Appl. Anal.* **2009**, *2009*, 135843. [CrossRef]
35. Kar, T.; Batabyal, A. Persistence and stability of a two prey one predator system. *Int. J. Eng. Sci. Technol.* **2010**, *2*, 174–190. [CrossRef]
36. Tian, X.; Xu, R. Global dynamics of a predator-prey system with holling type II functional response. *Nonlinear Anal. Model. Control* **2011**, *16*, 242–253.
37. Smith, H.L.; Thieme, H.R. *Dynamical Systems and Population Persistence*; Graduate Studies in Mathematics; AMS: Providence, RI, USA, 2011; Volume 118.
38. Alebraheem, J.; Abu-Hassan, Y. The Effects of Capture Efficiency on the Coexistence of a Predator in a Two Predators-One Prey Model. *J. Appl. Sci.* **2011**, *11*, 3717–3724. [CrossRef]
39. Alebraheem, J.; Abu-Hassan, Y. Persistence of Predators in a Two Predators-One Prey Model with Non-Periodic Solution. *J. Appl. Sci.* **2012**, *6*, 943–956.
40. Alebraheem, J.; Abu-Hassan, Y. Efficient Biomass Conversion and its Effect on the Existence of Predators in a Predator-Prey System. *Res. J. Appl. Sci.* **2013**, *8*, 286–295.

41. Alebraheem, J.; Abu-Hassan, Y. Dynamics of a two predator–one prey system. *Comput. Appl. Math.* **2014**, *33*, 767–780. [CrossRef]
42. Alebraheem, J. Fluctuations in interactions of prey predator systems. *Sci. Int.* **2016**, *28*, 2357–2362.
43. Hsu, S.B. On global stability of a predator-prey system. *Math. Biosci.* **1978**, *39*, 1–10. [CrossRef]
44. Ameixa, O.M.C.C.; Messelink, G.J.; Kindlmann, P. Nonlinearities Lead to Qualitative Differences in Population Dynamics of Predator-Prey Systems. *PLoS ONE* **2013**, *8*, e62530. [CrossRef] [PubMed]
45. Abu-Hasan, Y.; Alebraheem, J. Functional and Numerical Response in Prey-Predator System. *AIP Conf. Proc.* **2015**, *1651*, 3. [CrossRef]
46. Alebraheem, J.; Abu-Hassan, Y. Simulation of complex dynamical behaviour in prey predator model. In Proceedings of the 2012 International Conference on Statistics in Science, Business and Engineering, Langkawi, Malaysia, 10–12 September 2012.
47. Smout, S.; Asseburg, C.; Matthiopoulos, J.; Fernández, C.; Redpath, S.; Thirgood, S.; Harwood, J. The Functional Response of a Generalist Predator. *PLoS ONE* **2010**, *5*, e10761. [CrossRef] [PubMed]

© 2018 by the author. Licensee MDPI, Basel, Switzerland. This article is an open access article distributed under the terms and conditions of the Creative Commons Attribution (CC BY) license (http://creativecommons.org/licenses/by/4.0/).

MDPI
St. Alban-Anlage 66
4052 Basel
Switzerland
Tel. +41 61 683 77 34
Fax +41 61 302 89 18
www.mdpi.com

Symmetry Editorial Office
E-mail: symmetry@mdpi.com
www.mdpi.com/journal/symmetry